July 2003

Dear Nicki Ann

This book is about the place I spent
4 years of my childhood from 1934 to 1938.
The technical stuff wont mean much, but
the information about the people, a lot
of whom I knew, is interesting. To be
more precise, the author, Dean Herring
who I knew, was my boy Idol. I do.

This place has held beautiful memories
All my life.

Love,
Dad,

Kimberly, Utah
FROM LODE TO DUST

By
Dean F. Herring

HERRING PUBLISHERS
Eugene, Oregon

Library of Congress Catalog No. 88-91311
ISBN No. 0-9621854-0-X

Copies of this book are available from:

HERRING PUBLISHERS

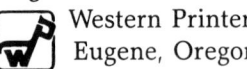

Mrs. Deloris Herring
625 Sterling Dr.
Eugene, OR 97404-2932

(503) 689-0920
541

Design and Production by:
Western Printers
Eugene, Oregon

Acknowledgments

I wish to extend thanks and acknowledgment to the many people whose enthusiasm and contributions kept this book alive. Among them:

Mr. Edward G. Tilton whose priceless documents, photos and endless stories gave body to this project.

My sister, Maizie Nielsen, for her enthusiasm, support, and tireless research of newspaper files.

The *Richfield Reaper* for their fine help and cooperation with the publication and publicity with the original edition of this book.

The Utah Historical Society for their *Historical Quarterly,* photos, and prompt response to inquiry.

The long defunct *Mining Review* on file at the Salt Lake City Library.

Elmo Herring for his notes and photos.

Charlie Hansen for his photos and stories.

Rehnon Nelson, Oluf Olsen, Woodruff Sylvester and Hamner Smith for their colorful stories.

Ken Lawson for use of the 1903 Kimberly school records which he had presence of mind to salvage and preserve from time- and scavenger-ravaged debris of the Kimberly schoolhouse.

My father, Floyd Herring, for his documents and stories, and for making it possible for me to spend (as Woodruff Sylvester so colorfully described) "the golden hours of youth there."

Once the center of a booming gold mining region, the town of Kimberly (outlined) leaves hardly a trace.

Contents

CHAPTER 1

A Legend Is Born

The giant mountain stirred from winter slumber and cast its cloud blanket leeward. The snow-capped peaks stood splendid and supreme, untouched by man, guarding its treasures in smug complacency. Snow crystals quivered under the caress of the morning sun, then melted away to form rivulets that trickled into networks of streams to the valley below. Downstream, a trout leaped for a deer fly, then plunged again into the icy water unheard above its cascading roar.

Farther downstream, freshly gnawed logs reinforced an imposing beaver dam. A nearby cluster of stumps carved to a point by chisel-like teeth attested to the seriousness of this community undertaking. Below the dam, the tumbling water, gathering in mass and teaming with trout, wound its way to the dry valley below.

Lush stands of timber carpeted the lower reaches of the mountain. The trees drank in the sun and beckoned nature's creatures to venture forth from sheltering branches. A doe and a fawn emerged from a dense growth of spruce and picked their way across a meadow. The doe's eyes darted from side to side seeking new shoots of edibles while watching the frolicking fawn. A porcupine ambled along the edge of a small creek, then disappeared into the dense underbrush. A noisy chipmunk scurried among the branches of a spruce, then scampered out onto the meadow to sun himself on a fallen tree.

This mountain looked no different than many others. Like most, it was a conglomerate of varied mineral, plant and wild life; exciting to the adventurer, perplexing to the weary traveler and refreshing to the casual viewer. But this mountain was different. Ominous outcroppings of quartz laced its slopes. here and there, bright yellow streaks of gold intersected the quartz, and like tiny veins, penetrated to the heart of the mountain.

Man's first excursions into the Gold Mountain area can only be conjecture. Indian petroglyphs, vast in quantity and rich in quality, adorn nearby canyon walls. Indian artifacts can still be found at several campsites among these towering Tushar Mountain peaks.

Standing in a clearing near one of these campsites in Surprise Gulch, one can imagine bustling life as it was then. A freshly killed deer hung in a nearby tree. An old man slapped at a deer fly as he chipped at a new arrowhead. Two women in adorned deerskin frocks scraped a freshly stretched hide. Children splashed and played in the crystal clear creek. Later, in the dusk of evening, campfires sent spiraling columns of smoke into the unblemished mountain sky.

This was a summer camp, because in winter, even the wild game sought refuge in the lower reaches of the mountain from the harsh deep snows. The icy peaks, though majestic in winter, offer little to the needs of man.

In the early 1800s, a new habitant began his encroachment into the lower valleys. Though much

Deep in the heart of Utah, the mighty Tushar Mountains reach for the sky gracing the earth with their majestic beauty. Photo by Dean F. Herring.

like the Indian, his skin was white, and his habits were vastly different. Unlike the nomadic man with red skin who drifted from camp to camp with seasons' change, the white man sank his roots into the rich earth like a weed extracting the wealth from the soil.

Villages sprang up, and as each new bastion was formed, more settlers arrived, and like water from an overflowing dam, they spread over the valley floors. Many staked out large tracts of farmland and tilled bountiful crops. Some trekked the streams and mountain trails in search of wild animal furs, and others plied the mountain streams in a never ending search for gold.

Names and dates of the first gold discovery on Gold Mountain have become clouded by time. Early stories tell of Indians hiking to the valley below carrying gold and spreading tales among the new white settlers citing the mountain's riches. One story tells of an Indian who brought a quartz sample showing flecks of gold to the village of Joseph. The white man gave him food and a promise of more if the story of the gold's location was true. They hiked to the mountain and found a narrow vein approximately two inches wide and of rather poor quality. But it was rich enough to start an influx of fortune hunters to the area, and soon it was found that the whole hillside would "pan." But again, the quality was only rich enough to whet the growing appetite for gold.

In 1868, on the east slope of the Tushar Range ten miles away, the Ohio Mining District was organized. These mines, known as the Bully Boy, the Morning Star, the Union and others, exploded into a beehive of activity almost overnight. Glowing reports of 13 ounces of gold and 200 ounces of silver per ton of ore from the Copper Belt mine raced through the mining world. By 1872, Bullion Canyon had 50 buildings and a population of hundreds. The entire east side of Gold Mountain crawled with prospectors, and tunnels appeared like molehills on every slope.

Some of the more exciting finds on the mountain included the Deer Trail Mine discovered by a Joseph Smith while trailing a wounded deer. The Webster mine, ignited by an enormous vein of lead and silver ore and fed by the fuel of lust and hope, sprang into a full blown community. Probably the most exciting single find ever made on the mountain was the Deer Park Boulder. Discovered by two of the hordes of prospectors panning for gold near the original gold discovery site, this huge rock reportedly weighed approximately 30 tons and assayed more than $160 per ton. It was found on a hillside one-half mile west of what was later to become the Sevier Mill site. The boulder was broken up and hauled by mules to the nearest road on Mill Creek, then shipped to Salt Lake for processing.

For every dollar realized from the sale of this

Gold Mountain looks no different than many others. But this mountain was different. Coursing through its monstrous bosom, yellow veins of gold ran wild.

giant rock, hundreds were spent in search of its source. Some speculated that it was an outcropping that had broken away from the mother lode, then been left standing while the main body sank into the earth's depths. Others believed that it had been deposited by an ancient glacier. Its origin has remained a mystery and continues to intrigue the modern-day prospector.

Little has been written, and very little is known of these early adventurers; however, the mole-like diggings over the miles of mountainsides bare evidence of the fever that once swept the area. The names of hundreds of claim holders fill dusty records in Junction, the Piute County seat. Most have long since been forgotten. Some staked their all against the confounding mountain. Some found riches, some found adventure, and some found heartbreak and death.

The scope of these diggings shed some light on the character of the men who worked them when viewed in the mountain ruggedness and rarified air of this 9,000 foot undertaking. Names such as Joe Jarrett, Joe Nelson, Newton Hill, Billy Johnston, Willard Snyder—names too numerous to

Petroglyphs adorn canyon walls below Kimberly.

Typical of the countless Gold Mountain mines was the Etheridge mine, on the Marysvale side in 1918.

mention are legend. The remains of their cabins and diggings can still be found on the dozens of hillsides from Kimberly to Sevier, south to the base of Signal Peak, and east to the Deer Creek and the Bully Boy. Hadly a square inch of mountain remains unclaimed. The names attached to the claims are as varied as the personalities of those who claimed them. Some had the ring of patriotism or nationalism such as the Statehood, The Britannia, The Royal George, The Dictator, The Columbia and The Pride of America. Birds were popular as evidenced by The Bluebird, The Red Bird, The Gold Bird, and The Snow Bird. Some indicated hope, such as The The Prosperity, The Climax, The Surprise and The Good Enough. And some registered desperation such as The Last Chance. Then, as now, the names of girls were prominent. The Geneva, The Lillian, The Hazel Kirk, The Jessica, The Minnie Maude and The Annie Laurie were but a few.

In the end, it was the Annie Laurie that stood out above all the rest. It was the rich lode of this mine that formed the nucleus for the town of Kimberly, and eventually, the name Allied Annie Laurie Gold Mines Inc. far overshadowed the multitudes of Gold Mountain claims.

Gold Mountain's environment is both delightful and harsh. Altitudes range from 5,300 feet at the valley floor to 12, 173 feet at the peak of Mt. Belknap, a zone of almost perpetual snow.

The settlement that became known as Kimberly nestled among peaks between the 8,000 and 9,000 foot level. The enchanting summer temperature seldom climbed to 90 degrees. The mostly cloudless skies allowed the summer sun to pour its life-giving warmth into the lush vegetation. The nights were cool, but not cold. Countless streams tumbled down ravines and across meadows. Green stands of pine and spruce shared the hillsides with soft trembling quaking aspen. Elderberries and choke cherries grew in abundance. Wild strawberries, raspberries and gooseberries, while not profuse, added to the richness of the mountain wilderness. Wild flowers—columbines, bluebells, and countless other varieties—abounded. Deer and small animals—porcupine, woodchuck, beaver, chipmunk and pine hen—were at home here. An occasional coyote, lynx and mountain lion were evident to the trained eye. The high altitude was above the range of the troublesome rattlesnake.

Nowhere is the wonder of nature more revealing, nor is the insignificance of man more distinct than hiking the trails through the timbered mountains and towering peaks. The body is tranquilized by the warm sun's gentle stroke, and invigorated by the cool rush of mountain air.

With the coming of winter, a new mood grips the mountain. The green leaves of summer turn brilliant yellow and fiery red, then wither to an ugly brown. Familiar landmarks are covered with snow. Food is scarce and icy winds whip the snow into drifts. Animals disappear into hibernation, or seek lower elevations. A new beauty emerges. The lofty peaks glisten in their new white blankets. The trees bend as though bowing in humble prayer under the weight of their heavy veil. The silence is awesome and the spectacle splendorous. In winter, here again, the insignificance of man is even more distinct.

It is awe and dismay that grips one then, to view such grandiose splendor raped by man. For man, of all earth's creatures, is gripped with the unquenchable desire for possessions and equipped with the intelligence to overcome resistance to his drives. It is the tragedy of man that his hunger for possessions blinds him to the necessity for preservation. Kimberly is a textbook on nature versus man. Man's drive for self preservation, his obsession for wealth and position, and his blatant disregard for his environment and heritage are all a part of Kimberly. The power of nature to control man and recover from his onslaught is displayed graphically on Gold Mountain. In the incredible short span of less than 100 years, a thriving town and industry, with all its hopes and

dreams, happiness and woes, sprang into existence, then vanished almost without a trace.

The great mineral belt of the Tushar Range is of volcanic origin and bends like a horseshoe around Mt. Baldy. Its 70 mile length is laced with gold and silver-bearing ore made of great porphyritic dykes. They cooled in ridges when they were formed with troughs between them. The veins formed near the apexes of the ridges or mountain spurs.

During the ice ages, glaciers moved down the troughs forming deep canyons, but leaving the gold almost untouched. Having no watershed above them, the veins remained almost in the identical conditions as when formed. A 1900 issue of the *Mining Review* stated, "These strong parallel veins are true fissures in porphyry, a condition which almost absolutely insures continuity and permanency as well as large, and practically inexhaustible ore bodies."

The grey dump at the 9,500-foot level (upper center) was the Annie Laurie #1. At this spot on June 21, 1891, Newton Hill placed his markers and filed claim giving birth to a legend. Number 2 Annie Laurie is hidden by a clump of trees below, and to the right. Bluebird group is on the left. Photo by Dean F. Herring.

CHAPTER 2

Exploration and Development

The discovery of gold ignited an inferno of gold fever. Diggings erupted like molehills on the hillsides. They began in the area known as Sevier and spread in all directions. The main interest was to the south on a hill later known as "Tip Top," then on into Fish Creek.

It was in 1891 that Newton Hill filed a claim that was to bear the most famous name in the area. This claim filed in the Junction courthouse reads as follows:

"Beginning at a broken-topped pine tree in an easterly direction from this monument, thence southeasterly 800 feet to a scaled tree, thence west southwesterly 600 feet to a stone monument, thence northwesterly 1500 feet to a scaled tree, thence east northeasterly 600 feet to a stone monument, thence to the place of beginning. The mining claim shall be known as the Annie Laurie mine. Located this 22nd day of June, 1891."

The steep hill climb to the Annie Laurie Mine was passable only to mules and man. There were no roads to the area. The formidable task of hauling ore was, at first, solved by sliding it down the hill on cowhides stretched between two poles. At the foot of the hill, it was loaded onto pack mules to be carried down the trail to the road.

The Golden Star Mine, later renamed The Bluebird, located 300 yards to the south of the Annie Laurie, attracted widespread attention even before the Annie Laurie, and later produced some of the richest ore to come from the mountain.

Gold and silver bearing ore is usually found relatively near the surface on Gold Mountain. It is rare that a vein extends extremely deep. Logically, then, it would be cheaper to dig a shaft down to the ore body. Unfortunately, due to the unusual amount of water, it was necessary to provide a means to evacuate it. Tunneling in from a lower level proved to be the most practical solution. This allowed the water to flow from the tunnel by gravity. Also, with the tunnel dug, removing the ore from the mine was much easier than raising it from a shaft.

For the above reason, then, when it was discovered that the vein of the Annie Laurie led downward, a new tunnel was dug about 100 yards down the hill from the original tunnel. It became known as Annie Laurie Number Two.

The Golden Star, or Bluebird to the south, developed similarly. Number Two Bluebird was dug 150 vertical feet below Number One. Then 300 feet below Number Two, the famous Number Three Bluebird came into being. By August 1900, the Bluebird had produced $22,000 in yellow metal. The ore often averaged eight to 12 ounces of gold and 20 ounces of silver per ton. Under the management of Pat Ryan, this mine gave employment to eight men.

Like a new formed anthill, the area buzzed with activity.

Fortune hunters swarmed to the area. The Snyder brothers, Willard and Greely, were among those who rushed to the mountain in 1896 gripped with gold fever, and obsessed with great expecta-

Number Four Annie Laurie became the most prominent mine of the Gold Mountain area. Photo, 1900, from the Mining Review.

The rush of prospectors and miners brought families to the mountain. Housing ranged from tents to four room bungalows. Photo, 1900, courtesy of Mrs. M. E. Jewkes.

Home of newlyweds William Avery and Martha Hansen in 1901. Avery worked on the water flume from Fish Creek to the power plant there. Photo courtesy of Mrs. Lincoln Avery.

Mr. and Mrs. William Avery at home. The bare necessities of life were generally the rule in Kimberly. Photo courtesy of Mrs. Lincoln Avery.

tions. They formed the nucleus of a small settlement along the east fork of Mill Creek approximately two miles by trail and 1,000 feet lower than the original Annie Laurie Mine. The village became known as Snyderville and was later to be named Kimberly.

Small businesses sprang up and more diggings erupted. A road inched its way up the mountain into what later became known as Upper Kimberly. Then a chance discovery set the stage for more rapid development. A decision was made to cut a road up the side of a hill to the Annie Laurie and Bluebird tunnels. The subsequent cut uncovered a rich vein of gold-bearing ore. This new mine became known as the Annie Laurie Number Three. Excitement ran wild. It was apparent that the whole mountain was laced with gold-bearing veins. A stamp mill was rushed to the site and a milling operation began.

Most of the digging was through solid rock. Strategically located holes driven by hand with sledge hammers and sharpened steel rods took their toll in human endurance in the 9,000 foot rarified air. Dynamite tamped into the holes and detonated by a fuse and a detonating cap reduced the rock to a manageable size. Muckers shoveled the fallen rock and muck into wheelbarrows or ore cars for the trek from the tunnels. Timbermen installed eight- to 12-inch diameter timber uprights along the progressing tunnels. An eight-by-eight-inch diameter "cap" crossed the top of the uprights, and "lagging" (usually two-inch thick boards) lined the overhead and walls to stop the falling rock.

Steam-powered air compressors found their way to the larger mines to run jack hammers for rock drilling. This modernization greatly boosted production. Before each shift, the miner was issued four candles. These were to last him for the entire shift. The flickering dim light in the pitch black tunnels barely lighted the working area and did little to illuminate the potential hazards. The candles were awkward to manage and constantly went out in the cramped working areas increasing the dangers. Hazardous working conditions, common to all mines, plagued workmen. Falling rock and cave-ins were constant threats.

On the plus side, the temperatures in the mines were ideal. Through summer and winter, the temperature varied only a few degrees. Compared to outside temperatures, it was warm in the winter and cool in the summer.

More claims developed, both in the area of Snyderville and the hills beyond. The Bluebird, the Grasshopper, the Hastings, the Surprise and countless others dotted the hillsides. The vein in Number Three Annie Laurie dipped lower and the ever-present water made it necessary to dig another tunnel to intersect the vein at a lower level. This new tunnel became Number Four Annie Laurie and was located one-quarter mile north of number three. This became the most important mine of the entire Gold Mountain development, and the focal point for the development of Upper Kimberly. The tunnel entered the mountain at the 8,450-foot level.

Willard Snyder's driving influence spread to Sevier. He staked a claim one-half mile northwest of the Sevier mill site and named it The Holland. The surface and outcroppings showed good promise and he was convinced that he could drive a tunnel below the Sevier mine and intercept the main vein. Being a promoter by nature, he attracted the interest of the Cannon family in Salt Lake City, who financed the project, and work began. Greely Snyder was hired as manager of the operation. He moved his wife to Sevier and she operated a boarding house.

The Holland Mine was an ambitious undertaking. It employed an average of 30 men and work continued for six years. The tunnel penetrated the mountain approximately one mile, much of it through solid rock. It passed through old glacial deposits which proved even worse than the rock. The soft muck shifted continually under the pressure of rock masses. The movement crushed the supporting timber like match sticks and repair work was constant. Steel posts and lagging were tried, but proved costly.

From time to time, glowing reports in newspapers and magazines acclaimed the vast riches of The Holland Mine. One such report in the *Mining Review* in 1903 stated that the Holland Mine ". . . has yielded the richest assays on the mountain. Laying on the dump of the shaft, in which work has been suspended, are tons of ore in which free gold can be seen with the naked eye. Recent investigation has satisfied owners that the vein can be opened in a more substantial form in another place." It was not explained why the work had been suspended in the "rich" shaft, nor why the "rich" ore was not being shipped to the mill. Reports such as these did attract investors, however, and investors were a badly needed commodity.

In September 1903, the Holland and Deer Creek properties were consolidated; the Deer Creek going to the Holland for $20,000 to be paid in 20 months. This enabled the Holland to continue its tunnel project which was nearing Deer Creek property.

In the end, money ran out and the rainbow faded. The mine closed without having shipped a pound of ore. This was a prime example of hope, hazard and heartbreak in the search for gold. It has been concluded by some that the Snyder brothers worked against two theories. They were attempting to tap an ore body below another claim. If they had reached the ore, the cost of recovering it from the long tunnel and transporting it to the mill would have been considerable. Also, they drove very deep into the mountain. In the Gold Mountain area, the richest ore bodies are near the surface. Thus, they were defeated before they began.

In about 1910, a cloudburst sent a cascade of water and mud to bury the sheds, the mouth of the tunnel, a small steam engine, the steam boiler and fan at the Holland. As late as 1922, the upper ten feet of the smoke stack still protruded through the muck from the slide.

Meanwhile, over the hill in Snyderville, it

Isolated in the bottom of a canyon, the old Holland mine cabin remained relatively intact in 1974.

became apparent that the crude makeshift mills were inadequate. Willard Snyder, although busy with the Holland Mine in Sevier, formed a small company for the purpose of developing a modern mill. About 1899, he contacted and hired a man by the name of Ephraim Clapp in St. Louis, Missouri, a man with some engineering ability to head the new operation. Clapp also guided Snyder to a syndicate in Sharon, Pennsylvania headed by Peter L. Kimberely and L.C. Huck. In April 1900, after several weeks of negotiations, the Annie Laurie Gold Mines were sold to Kimberely for $210,000. Great credit was given to the "inde-fatigable" efforts of Willard Snyder for this accomplishment. The property of the Snyder Improvement Co. was included in the deal.

Under Clapp's direction, Snyderville became "Kimberly." (No explanation has been found for the difference in the Kimberley versus Kimberly spelling.) A corporation was formed and stock sold. A large mill was built on Mill Creek at the 8,100-foot level. This construction required large logs not available on Gold Mountain. These were shipped in from out of state. The timber was cut and framed at the railroad station to avoid shipping costs and waste.

A new tunnel was begun at the new mill site. Called Number Five Annie Laurie, it was located approximately one-half mile north and down the hill from Number Four tunnel.

The new mill originally had a capacity of 200 tons per day. When it began operation in February 1901, publications claimed that no plant was ever constructed on better economic lines. At that time, the Annie Laurie was estimated to have $3,000,000 net reserves, the ore running $10 to $20 per ton. The *Mining Review* stated, "Of all Utah's low grade gold mining propositions, the average gold content is the highest in this mine."

The method of ore treatment in the new mill consisted of coarse crushing, then drying the ore in a huge tubular roaster. Conveyors carried the dried ore to the fine rollers where it was crushed into a fine sand, then classified with impact screens. From there it went into enormous tanks for a process known as cyanide leaching, or cyanidation. In this process, the gold is dissolved by the cyanide solution, then percolated from the solution by zinc dust, or electrolysis.

Ten 150-ton and six 300-ton tanks occupied a long low extension to the north side of the mill for the cyanidation process. The larger tanks were 30 feet in diameter and approximately six feet deep. A solution of eight to ten pounds of cyanide to one ton of processed ore remained in the tanks from eight to ten days.

Mrs. Josephine Pace, a late resident of Richfield, Utah told of her girlhood experiences in Kimberly. She described watching the crews of men "with long poles tending, stirring the thick choking liquid. Masks covered their noses and mouths, and arm high rubber boots disguised them." To this

The large Annie Laurie mill on right was obsolete before it was completed. Wood from the denuded hills lays stockpiled (center) for drying ore in huge tubular roasters. Tramway from mill to Upper Kimberly to right of stacked wood. Lodge, extreme left center. Photo courtesy of Utah Historical Society.

young girl, "they looked like men from never, never land."

In reality, what appeared to be "a thick choking liquid" was a five-tenths percent cyanide solution which actually emitted no gases. The masks were worn for protection from dust and the high rubber boots protected against splashing.

At the end of ten days, the solution was changed to one-half strength. Before cyanidation, the tailings material was sluiced to the plate room. This room was so named because of large copper plates covered with mercury over which the treated pulp was passed. The fine gold particles, being heavy, settled into the mercury forming an alloy. The mercury was then vaporized by heat leaving the gold intact. This process was called amalgamation.

The tragedy of the mill was that it was obsolete before it was completed. From the standpoint of metal recovery, the mill was very thorough; 93.4 percent of the gold and 40 percent of the silver was recovered. A high percentage of manganese interfered with silver recovery. The processed ore varied from $5 to $200 per ton. The average assay reports showed approximately one part gold to four parts silver. From 1901 to 1908, the mill processed 425,000 tons of ore and had a bullion recovery of $2,785,000, or $6.55 per ton. About $1 per ton was in silver. From September 2, 1902, until April 22, 1905, the company paid dividends amounting to $439,561.

The gold was melted into bricks ranging in size from five to 25 pounds. They were carefully weighed, then sampled. Sampling consisted of drilling a small hole in the brick and assaying the drillings. By knowing the exact percentage of metals in the brick, a precise value could be placed upon it immediately, which was important for insurance purposes. The shipments were insured until their arrival in Salt Lake City.

From a standpoint of efficiency, the mill was a colossal failure. There were 13 elevators for transferring ore and the capacity was seriously impeded by mechanical troubles. One of the primary flaws, however, and one of the reasons for the eventual collapse of the company, was the drying process. Sixty percent of the cost of the milling operation was in supplying fuel for drying the ore. The huge revolving dryers consumed 80 cords of wood per day. The nearest coal supplies were far beyond the range of available transportation. Heating oil had not yet entered the industrial scene

in the area. Wood was plentiful—or so it seemed. Crews of woodcutters spread out over the mountainsides. The area near the mill was soon stripped of vegetation. The distances to new supplies lengthened, and accordingly, the crews grew larger. At its peak, estimates placed the number of men in the wood crews alone at near 300. Three large chutes on the slopes of the east mountain gravity fed wood and logs to waiting wagons below. At least one man was reported killed when a hurtling log bounced from the chute and struck him.

For these pioneers of industry, ecology was only a word in the dictionary. The denuded hills and the garbage disposal were prime examples of environmental disregard. The environmental impact of the huge tailings pond, locally referred to as the cyanide dam, has never been officially assessed and remains a subject of contradictory views. This imposing dump consisted of very fine sand—the tailings from the mill after the gold and silver had been removed—and unquestionably had been exposed to cyanide and mercury. Except for occasional accidents, it is doubtful that cyanide content ever reach perilous levels. In one instance, a new "cyanide man" increased the strength of the solution to ten times the normal strength. Three or four cows grazing along the creek died when they drank the water. It is a fact, also, that signs posted on Mill Creek warned against drinking the water because of cyanide.

Mercury, an insidious killer, was not fully understood at the time of Kimberly's boom. Only in recent years has its ability to contaminate water supplies and our food chain been brought to light. Without question, large quantities of mercury found its way into the tailing and into the stream. It is commonly known that fish were virtually nonexistent in Mill Creek until recent years. Some argue that there were never any fish in Mill Creek even before the mill's existence. Since fish are abundant in other streams in the area, the argument can be made that the contaminants killed them. The tailings dump covered several acres one-quarter mile below the mill, and in places was over 50 feet deep. Rain and snow water leached, and after 80 years, continues to leach the sand into Mill Creek which feeds into the Sevier River.

The town of Kimberly sprang up almost overnight. Workmen with families swarmed to the mountain. It began with a few tents and shacks

Huge tailings dump, locally referred to as the cyanide dam, gives way to new growth as persistent aspen push their roots into the deep sand. Photo, 1978, by Dean F. Herring.

hurriedly thrown together from logs and scrap lumber. Some tents had a lean-to attached to serve as a kitchen or additional bedroom. As the town's stability grew, more permanent housing appeared. Businesses blossomed and boomed in Lower Kimberly, about one-half mile east of the mill on Grasshopper Creek. A small steam-powered sawmill was built a few hundred yards up the creek to the south of Lower Kimberly. This supplied the much needed lumber for the mushrooming community.

Fire was a common threat. Inadequate chimneys were the rule rather than the exception. Hot stove pipes ran through the walls or ceilngs which were often held away from the tinder dry wood or canvas by only a sheet-metal shroud. There was no fire department. Water pressure was low. Fire hydrants were non-existent. One story tells of a large stable fire. The town turned out to fight it. They released 30 horses, but the pigs were another matter. They buried themselves in the straw and had to be bodily forced outside.

By now, Kimberly was truly a roaring boomtown. One writer of the time wrote, "We have the ore, plenty of water and timber, the Rio Grande Western Railroad and the greatest scenic attraction on earth. If the mining fraternity wants anything more, we can find it."

By 1900, daily stage service to the railroad at Vaca linked Kimberly with the outside world. A writer for the *Mining Review* wrote, "Fourteen

miles of excellent wagon road winds its way up Clear Creek Canyon, the scenery along the way being magnificent, picturesque, and rivaling the Grand Canyon, almost, in its scenic beauty. In Kimberly, on every hand, the sound of the hammer and saw is heard. New and still unpainted buildings attest to the rapidity and impatience with which the contour of the country is being changed from an almost un-penetrable forest, into the abiding place of man. Re-echoing blasts coming from the mountain sides indicate too plainly that every effort is being made to penetrate into the secret depositories of nature in the persistent search for Wall Street metal; for gold which enriches the world, in the recovery of which no man is robbed or injured." Although this writer's description seems overly dramatic, it captures the fever that existed.

In 1900, the extension of the Rio Grande Western Railroad to Marysvale added new zest to the exploding mountain. A wagon road from Kimberly to Marysvale, completed in 1900, linked the area even closer to Salt Lake City and the east.

In August 1900, another writer for the *Mining Review* wrote, "Expectations are now being realized. Neglected for years, save to the hardy and persistent claim-holders who realized what was there, large capital needed in the development and operation of the great fissure veins is now being freely and intelligently spent."

Boosters claimed that the consolidation of the

Annie Laurie Gold Mines would soon place Kimberly in the front ranks of the greatest gold producing sections of the West.

The name of Charles Skougaard is synonymous with early Kimberly. His talents extended to most areas of building contracting and construction. He built the majority of the permanent housing units in Kimberly. It was he who built the lodge, the most imposing building in Kimberly aside from the mill. He also built and managed a hotel in Lower Kimberly.

Electric power came of age, and now, Kimberly, too, was becoming modernized. A power plant in the bottom of Fish Creek soon became over burdened. In the spring of 1902, 40 workmen under the direction of Charles Skougaard began work on a canal and flume from the head of Fish Creek near Signal Peak to a new power plant one-half mile north of the overworked facility. A 12-inch riveted steel header pipe directed the water from the flume down the hill to the new plant. Mrs. Josephine Pace, daughter of Mr. Skougaard, told of her recollection of the first test of the new plant:

"Everything was finished, and the time had come for the first test. Father warned everyone to keep an eye on the pressure gauge. The water filled the pipes, the gauge began to rise, the wheels began to turn. Then, as Dad told it, 'Some dang fool rang the dinner bell and the men all ran down to camp.' Someone threw the switch back, and the water force ripped the pipes out with a roar."

In Clear Creek Canyon eight miles away, a new and bigger plant began to take form. Built by the Sevier Consolidated Gold Mines Inc., it was designed to operate on either steam or water power. The steam plant, powered by coal, was completed first. It consisted of three return tube boilers arranged so that they could be used independently as needed. There was a ten-horse vertical engine and a tandem engine to run the large generators. A partially completed dam and pipeline to run the 250 horsepower generators were never completed.

Problems, both political and logistic in completing the power plant in Clear Creek Canyon, caused another steam plant to be built in Vaca. The 6,600-volt line from this plant linked with those in Fish Creek to furnish the much needed power for the mills. Besides the obvious value of supplying power, the company hoped these projects would provide a stable appearance for the operation and attract more money.

Construction boomed. Miners clamored for timbers and lagging for the rapidly growing tunnels throughout the area. A new sawmill went into operation near the entrance to Number Four Annie Laurie tunnel. A combination blacksmith shop and boiler room for the sawmill added to the growing complex.

Logging crews stripped the mountains around Kimberly and beyond, and the supplies dwindled. Two more sawmills joined the Fish Creek complex.

Early stamp mill near Marysville about 1900.

*First passenger train into Marysvale.
Photo, 1900.*

*The power plant in Clear Creek Canyon designed to operate from either
steam or water power, failed to reach full utilization. The water flume
was never completed. Transportation of coal limited its capabilities.
Photo, 1920, Ed Tilton.*

*The Oluf Olsen stage arriving in Kimberly
about 1903. Its arrival was the highlight
of the day. "We were never late for its arrival," Josephine Pace remembered. Photo by
Alma Anderson.*

Towering fir trees stand naked among the company owned houses on the stripped landscape. A load of wood winds its way down the steep road through Upper Kimberly to the mill below. Photo, about 1901, courtesy Utah Historical Society.

In June 1900, Charles Skourgaard added a planing and grooving mill to one of these operations.

Four- and six-horse teams dripping with lather strained at the heavy lumber wagons in the steep two-hour climb from the bottom of Fish Creek Canyon to the top of Tip-Top Mountain, then wound their way down to Kimberly. The chop of the logger's axe echoed on both sides of Tip-Top. Logging roads scarred the slopes through the slashed timber, and lush green gave way to an ugly brown.

The thirst for sudden fortune ran unabated. Dedicated men and opportunists alike sank their roots and staked their all in the Tushars. The mountain became their life and they were a part of the mountain. Aside from the Annie Laurie, operations ranged from the lonely prospector to the 100-man complex at the Sevier Mines.

The Sevier Mines Inc. was located three miles over the mountain northwest of Kimberly. This was the site of the original gold discovery in the area. At times, it was predicted that the Sevier Mines would rival Kimberly in size and value.

The setting of Sevier fired enchantment. An early writer's impressions are described in the *Mining Review:* "It's high upon the mountain, and from a vantage point on Signal Peak, the Sevier Valley in all its splendor spreads out like a panorama to the east and north, while to the south, Mt. Baldy and Belknap, in their imperial majesty, look down upon a region holding the wealth of Croesus within its bosom. Forests of Quaking Aspen and stately pines abound on the hillsides and in gulches, while rippling streams tumble down ravines and canyons."

The Sevier Mines Inc. was acknowledged to be one of the most promising gold properties in the state, or the West, for that matter. The driving force in Sevier was Charles Lammersdorf. Born in Germany in 1834, he came to New York in 1860. In 1872, he settled in Utah. He was described as always in the front of mining ventures, and he shed no tears over losses. Always hopeful for the future,

Charles Lammersdorf, born in Germany in 1834, was the driving force in Sevier. Lauded for his generosity, he was hopeful for the future. He was one of the few men to leave the mountain a wealthy man. Photo, Mining Review, about 1903.

he was lauded for his generosity and was always ready to "grub stake" his fellow prospectors. He was confident of an eventual "strike." His philosophy paid off and he found himself associated with the Sevier Mines.

The first crude mill in Sevier was housed in a small log structure. Its first short runs gave returns of $10,000. This further sparked Lammersdorf's enthusiasm and left him dissatisfied with his role as a co-owner. He labored until he was sole owner. He formed a new company and sold approximately $400,000 worth of stock. He enlarged the mill to accommodate ten stamps.

The Sevier group consisted of seven patented claims and two mill sites. It enjoyed the distinction of being the first patented ground in the district.

The ore body was described as "massive." Two ledges 150 feet apart were portrayed as "like sentinels to challenge the treasure seekers, and dipped down into the earth to untold depths." The ore vein in the ledges was ten to 30 feet in width and assayed from $12 per ton to $540. Some ore from Sevier shipped to Salt Lake prior to the erection of the mill netted the company $800 per ton after expenses.

By August 1900 the company was incorporated with 250,000 shares at $5 per share. Charles Lammersdorf held 242,000 shares. Sevier boasted of

Built under the direction of Charles Lammersdorf, this mill in Sevier produced first short runs of $10,000. It was later enlarged to accommodate 10 stamps. Photo about 1900 from Mining Review.

The mill at the Sevier Mines. Completed in 1903, this new mill with a capacity of 75 tons per day, was much smaller than the 250-ton per day mill at Kimberly. Using a new wet process, this mill was much more efficient than the Kimberly mill. Photo courtesy of Ed Tilton.

a boarding house, several bunk houses, an assay office, a blacksmith shop, and many dwellings. It was estimated that there was enough ore in its mines to last 100 years.

In 1902, money problems caused an assessment of one percent on capital stock. By 1903, the mill had reached its potential and further growth was dependent upon further cash outlays for a new mill. Charles Lammersdorf, whether by foresight or due to old age, sold much of his Sevier holdings for a purchase price of $141,000. He retained a considerable block of stock.

The new company, under the name of the Sevier Consolidated Gold Mines, came under the management of Captain C.H. Lawrence, an experienced mining engineer. A new mill sprang up under his guidance. It used a wet process known as the "Trent System" which eliminated the cost of drying the ore. Both the mill and the mining operation were designed for efficiency. All movement of ore, where possible, was by gravity.

The trail to the Sevier Mines. Photo about 1903—Mining Review.

A mine at Sevier about 1903. Photo courtesy of Utah Historical Society.

At the Sevier Mines about 1903. Photo courtesy of Charlie Hansen.

Chutes carried the ore from level to level in the mine, and finally, to the waiting ore car. The receiving level of the mill was below the tunnel level and the ore fed automatically into one of three large crushers. Conveyor belts carried the crushed ore to the stamp mills. The mill boasted of 18 1,000-pound stamps.

The next stop was an efficient amalgamation room. Here the crushed ore pulp flowed into the huge cyanide tanks for cyanidation. This process varied somewhat from that of Kimberly. The ore at Kimberly was ground finer than at Sevier; therefore the 18-foot wide and 20-foot deep tanks at Sevier were considerably deeper than the 30-foot wide, six-foot deep tanks at Kimberly.

The Sevier Mill became operational in October 1903. In November 1903, Captain Lawrence went to Denver to consult with the Mercur Mines concerning the new "Slimes" process. He hoped this development could extract more of the value from the tailings. Loss in tailings was reportedly running $20 per ton.

In October 1904, Captain Lawrence reported on the start up of the new mill. He stated that the new Erie 75-horsepower engine "could slip every belt in the mill, and have horsepower to spare."

Though much more efficient than the 250-ton-per-day capacity Kimberly mill, the Sevier Mill was much smaller with a capacity of 75 tons per day. Even so, the Sevier exceeded all expectations. The company had little trouble in raising more money for the purchase of equipment and for acquiring additional claims.

In October 1904, the Sevier Number Three Tunnel encountered Number Two Tunnel, an event described as "most gratifying." The planned connection reportedly did not miss the calculations by more than one foot. The ore body was described as "astonishing." The vein between the two tunnels was described as "at least 100 feet in width, and much will go $75 to $80 per ton. Some will run into the thousands." It was predicted that the mine would be able to supply almost unlimited ore.

A new tunnel started 1,500 feet below the mill was planned to run at least one mile under Tip-Top Mountain. Captain Lawrence predicted that the Sevier would be one of the biggest mining operations known.

The success of the Annie Laurie sparked a rush of promoters and new ventures on the mountain.

Several were incorporated in St. Louis, Missouri, and sold stock in the East. Some were well-intentioned, founded on faith and high hopes. Some were promotional schemes with one thought in mind: to fleece the stockholders. Of the latter, it was estimated that less than ten cents of every dollar raised was spent in exploring. The rest went into the promoter's pockets.

The Annie Laurie Extension was one of these new ventures. Located almost due south of the Kimberly vein, it was in no way connected with the Annie Laurie Gold Mines. Stock was sold by mail order. The exact figures of stock sold is unknown, but in April of 1903, capitalization was increased to $1,000,000, having a par value of $1 per share. Promoters claimed the raise was justified by recent developments and by the need to acquire territory. Stockholders were sold on the fact that the northern line of the Annie Laurie Extension group was only 1,200 feet from that "great dividend payer, the Annie Laurie Gold Mines." The sales pitch claimed that 3,500 feet of tunnel had been mapped out, and that upon completion of the work, management was confident that they would be working in the famous Annie Laurie vein.

Stockholders were exuberant in August 1903 with reports that the tunnel had struck vein matter. It was conceded, however, that the vein matter was only one inch in width, "but carries good value in copper." The find was taken as a good omen and work continued.

In November 1903, stockholders were gripped in great expectancy. They were told that the tunnel had reached a point where it was expected to intercept a substantial vein. "Indications are that the judgment of those developing the company would be vindicated," they were told. The only trouble was they needed more money. In literature published to promote the sale of stock, they described the blower to be installed at the mine as the latest in ventilation and compressed-air equipment. They stated that this would enable them to use air drills to drive the tunnel. In reality, the blower was only capable of blowing fresh air and was not a compressor. It was driven by a worn-out steam engine that was dragged to the property on skids. There were no roads to the property. Due to travel hardships and the distance involved between the stockholders and the isolated mine, the operators apparently felt quite safe from prying investigation.

For several years after the collapse of the Annie Laurie Extension, the Annie Laurie Gold Mines was plagued with letters from investors inquiring about their "guaranteed stock" in the defunct company. Most were from small investors with from $100 to $700 in stock.

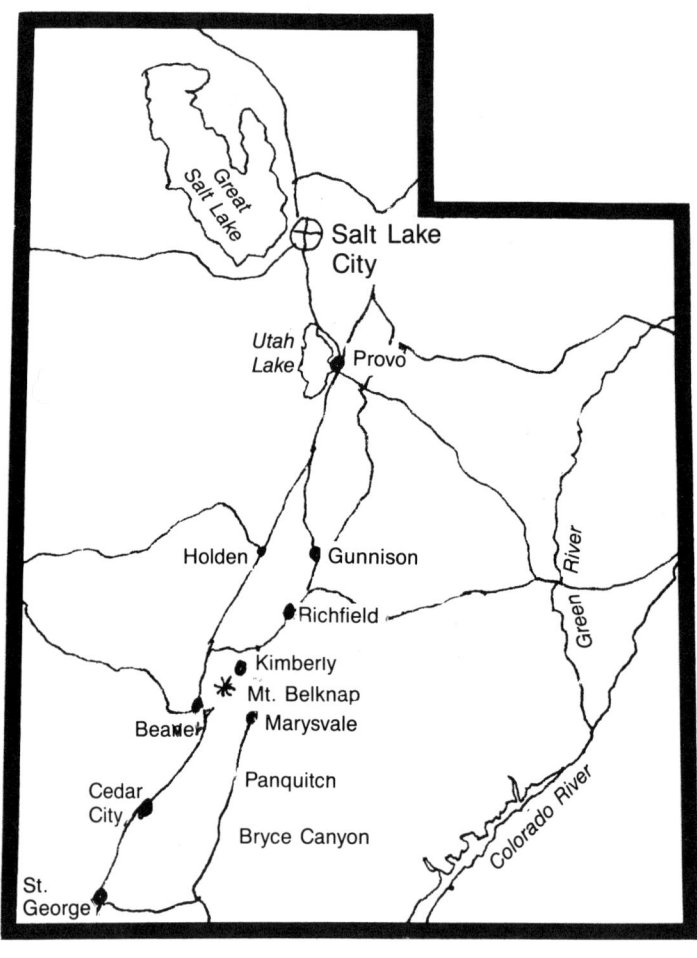

CHAPTER 3

Other Prospectors on the Mountain

Individual enterprise, the foundation of the "American Dream" fanned out across the mountain. Undaunted by failure, prospectors moved on to new diggings. To simply survive in the mountain wilderness was a challenge. Each claim was an outpost dependent only upon itself. From three to six feet of snow covered the hills six to seven months of the year. Firewood and food supplies to last through the long winters were stockpiled before the first snow. Most houses and cabins had "dugouts" or cellars to store food supplies. These were warm in winter and cool in summer. Kerosene lamps and lanterns provided light. Water supplies were no problem. Dozens of streams tumbled down hillsides and, in winter, if the stream could not be reached under the deep snow, there was always the snow itself.

The mountain seethed with prospecting activity from the Deer Trail Mine near Marysvale to the Signal Peak group at the head of Fish Creek. Brice Prisby, Jack Gilbert, A. J. Moore, the Darzer Brothers, Frank Shepard, Dwight Meteer, H. W. Ramloose, Walt James—and other names too numerous to mention, pitted their strength and ingenuity against the mountain.

Joe Nelson was typical of this hardy breed and his name is legendary. Among the early prospectors of the area, he combed hundreds of acres on Gold Mountain and built cabins on at least two of his many claims. His most well-known campsite was at the head of Fish Creek almost at the very foot of Signal Peak. Accessible only by trail,

the seven-mile hike to and from Kimberly tortured the lungs at this 9,500-foot level. The two single-room cabins on this site sat about 50 feet apart. The water supply was carried from a rock-lined hole in the creek that ran past the door. Two hundred yards below the cabins, Nelson's well-known Signal Peak Mine tunneled into the mountain. A sturdy log blacksmith shop housed the entrance to the mine, and steel mine car tracks from the tunnel led through the building and out onto the dump.

Like the majority of prospectors, Joe Nelson built most of the cabin, furnishings and other conveniences from the trees and material at hand. He fabricated an ingenious clotheswashing machine powered by water from the creek. Years later, in 1924, Floyd Herring was herding sheep in the area and was intrigued by the still intact machine. He went back to his camp, gathered his laundry, and hiked back to the cabin. After a few minor repairs, the old machine churned away with the load of laundry.

This new-found convenience, ahead of its time, lacked refinement, however, as Herring found out. Rather than wait for the laundry, he utilized his time rounding up sheep and promptly forgot the machine. When he returned a few days later, the water had leaked out and the threadbare clothing was still threshing around in the machine.

Nelson moved his family into the cabin and here, miles from supplies and human help, they burrowed in for the winter. Through the days, he drove his tunnel ever deeper into the mountain. It eventually penetrated 800 feet. During the long evenings, he played with the children, repaired equipment, or read. He was an accomplished musician and spent many hours playing his cornet.

The Nelson cabin was structurally superb. One severe winter, this fact saved his family when a snow slide buried the cabin. Only luck or some greater force pervented tragedy. The cabin was built in a swale, and the full force of the slide was absorbed by a small rise before striking the cabin. A son, Rehnon Nelson, remembered the incident. "I was only five years old at the time," he said. "My father was in the mine and my mother, my baby sister and I were alone in the cabin. We heard a big roar and the door burst open. Snow shot across the floor partially filling the room. The doors and windows were blocked. I remember mother looking for something to dig with. A pie tin was the only thing she could find and she started digging. The cabin was buried under 26 feet of snow. Dad came out of the mine and saw what had happened and began digging. They met each other about half-way through the avalanche. The snow and ice stayed around the cabin all the next summer and until August of the following year."

Joe and his brother, Chris, formed a band and traveled over much of the state. They built opera

An old bellows lays across the rails leading from the Joe Nelson Mine. 1978 photo by Dean F. Herring.

Joe Nelson's cabins built at the turn of the century at the base of Signal Peak still stand. One winter, about 1905, a snow slide buried the cabin under 26 feet of snow. Mrs. Nelson and her two children, alone in the cabin at the time, survived. Photo, 1978, by Dean F. Herring.

Joe Nelson's blacksmith shop at the entrance to his mine at the base of Signal Peak. An ore car with wheels missing lays upside down outside the shop. The rails leading from the tunnel lay twisted across the dump, then trail off into oblivion. Photo by Dean F. Herring, 1978.

The Oro Chain Mine of the Mammoth group taken 1900. Opened and operated by James Long, the mine boasted of some of the highest grade ore on the mountain. Although some good ore was taken from this mine, most of the claims were highly exaggerated.

Cabin at the Oro Chain Mine, 1977. Photo by Dean F. Herring.

houses including the ones at Kimberly, Junction and Richfield.

In addition to his many other interests, Nelson kept up the assessment work on several mining claims bought out by the Annie Laurie. In his own mines, however, gold eluded him. The ore values of his mines seldom exceed $5 per ton, too low to warrant processing.

In September 1903, Nelson incorporated four of his claims into Signal Peak, Inc. Combined with his Joe Nelson Number 3, 4, 5 and 6 was the Yankee Girl, the Golden Gate, Signal Peak 1 and 2, and the April Fool. It was incorporated with a capital stock of $30,000 divided into 300,000 shares.

In 1978, an old ore car, still intact, sat near the tunnel entrance. The entrance was caved in, but the steel tracks leading to the dump marked the spot that once inspired the hopes of fortune. A weathered grinding wheel laid half buried in the hard packed earth, and the blacksmith bellows lay propped against a log. Bullet shells driven into a log at the end of the cabin spelled the name, "Rehnon Nelson."

The Mammoth Mine, 1977. The log portion of the old bunk house stands stalwart amid the ruins of less sturdy structures. Photo by Dean F. Herring.

An entanglement of twisted grey boards and timbers mark the remains of the ore chute from the Long Tunnel at the Mammoth. Photo, 1977, by Dean F. Herring.

With the spread of gold fever, glowing optimism gushed from each claim on the mountain. Some dreamers were sparked by genuine hope, and some contrived to promote financial investment. In 1899, the Deer Trail Mine boasted 140,000 tons of ore averaging $7 in gold and ten ounces of silver per ton. The Mammoth Mine located one mile west of Kimberly was owned by the Piute Investment Company of New York. It included sixteen claims located in the Mammoth Basin at the head of Mill Creek. Work eventually settled on two main tunnels, the Breckenridge (later called the Oro Chain), and The Long Cross Cut was driven to tap a vein at a vertical depth of 600 feet and

penetrated the mountain approximately 1200 feet. Water from the tunnel was turned to an advantage to operate two water wheels for power. Heavy timber covered the property, and plentiful springs and creeks provided unlimited water. The mine was opened and developed under the direction of James Long, a veteran operator on the mountain. By August 1900, the Mammoth complex included a two-story building which housed a business office, an assay office, and a dwelling for the manager and his assistant. It covered a ground space of 24 by 22 feet. A two-story 28- by 32-foot boarding house provided lodging for several workers. Nearby a bunkhouse with a separate

Workmen move a steam engine from a sawmill at the Trappers Pride Mine in Fish Creek. 1924 photo by Floyd Herring.

reading room housed additional help. Stables, barnyards, blacksmith shops and small outbuildings added utility to the property.

Reports from The Long Cross Cut Mine further fueled the gold fever fire. Claims of "strong outcroppings of quartz three to five feet in width profusely plastered with precious metal assaying $6,900 in gold, and several ounces of silver per ton" reached periodicals. The truth of these claims cannot be substantiated, and a report in 1903 painted a strikingly different picture. It stated that a new tunnel had encountered "a fine vein of the prized Annie Laurie character." The average Annie Laurie ore seldom assayed as much as $100 per ton. The general tone of excitement and optimism was no less glowing, however. The *Mining Review* proclaimed that a great future was predicted for the Mammoth "and with little additional expense, it is in a position to become one of the great bullion producers of the mountain."

In September 1903, a story of a freakish severed vein in the main tunnel sought to explain the failure to find fortune.

The Surprise Mine, located three-quarters of a mile west of the Annie Laurie Mill, showed great promise. In 1903, developers released encouraging reports, then conceded that the company needed capital to run more tunnel.

On the Fish Creek side, the Trapper's Pride, located near the bottom of Fish Creek, claimed its share of publicity. Originally operated by a Mr. Peterson, it worked a crew of 12 men and boasted an excellent boarding house. The vein was described as "eight feet wide and very rich" and was characterized the same as that of the Sevier Mine and "just as extensive."

In 1903, A. D. McLain became manager of the Trapper's Pride. It was claimed that the property had "unfortunately been operated by managers who had sunk fortunes into useless, and unnecessay work." The main tunnel was poorly timbered and had caved in in many places. McLain reopened and retimbrered the 966-foot tunnel and promised that, as soon as showing justified, a mill would be built. In October 1903, he announced that, "the big vein has been encountered and the

Collapsed under the snows of many winters, this cabin at the Trappers Pride Mine is typical of many on Gold Mountain. Photo by Dean F. Herring, 1978.

ore is of fine character." He claimed that this gave the property a large body of developed vein matter and stated that the Trapper's Pride was getting into position to be a big mine.

In 1903, the Keystone resumed operation after a long layoff. Management's intention was to try to tap the Trapper's Pride vein.

The Lone Tree Company under the management of H. W. Ramlose began building cabins for its work force and announced they would try to tap the Sevier ledge. Developments at the Trapper's Pride and others spurred the owners to employ a force of men and a 200-foot tunnel was planned.

The Advocate consisted of six claims and was claimed without question to have developed an "immense ore body." Two tunnels 300 feet in length penetrated the mountain. Two shafts 70 feet and 20 feet respectively sank downward to veins described as nine to 15 feet wide and running $5 to $20 per ton.

The Shamrock, a new company in 1903, adjoined the Advocate on the south. A 500-foot tunnel was claimed to have tapped a good vein. One of the owners, a Mr. Christensen, began forming a corporation for development of the property.

The Mt. Baldy Mine management announced a "good strike" in its 400-foot tunnel, and a spokesman stated that an investigating party "has made an offer to buy all treasury stock."

An air of intrigue surrounded a report of two executives of the Boston Gold Mining Company who rushed to New York "owing to some deficiencies among the directorate of the company, the stockholders are refusing to carry the company further." The report stated that the trip was to quiet suspicions and to make arrangements for continuance of development. "They are prepared to show there is every reason to believe that it has as promising a proposition as exists on the mountain." The tunnel was described as running on a vein of good ore that warranted development, but it "is a bigger and richer proposition that the company aims to uncover." According to the report, the two executives hoped to meet with the demoralized stockholders and restore confidence in order to secure contributions for further development.

The Miller group, located above and north of the Advocate, was owned and operated by J. T. Miller, a native of Pennsylvania. The vein was described as "the same strong quartzite ore as the Advocate." Miller was 59 when he arrived on the mountain in 1893. He, like most early prospectors, spent more of himself in hard labor and hardship than he extracted from the mountain's treasures. Miller lived contentedly, however, in a small cabin just west of the Sevier Mine. He drank of the mountain's beauty and added to its human riches. He was described as "kind-hearted and honest to a fault." In 1900, he sold his holdings to the ever-growing Annie Laurie Gold Mines.

Garth Herring, lifelong familiar figure on Gold Mountain, at Trappers Pride boarding house. Photo, 1956, by Floyd Herring.

The isolated Trappers Pride boarding house still stands relatively intact. Photo, 1978, by Dean F. Herring.

To the east of Kimberly in 1903, the Silver King and the Bank of England made flagrant claims. "This section of Gold Mountain is destined, without doubt, to be a greater district than Kimberly," they stated. "There are ideal spots for townsites, abundance of water and timber, and good mill locations. The area is so extensive that it will be a camp of much greater extent and importance than the Kimberly side."

The Silver Queen management was thought to be more honest than many. Located in a canyon east of Kimberly, they drilled approximately 4000 feet of tunnel, built a large boardinghouse, housing for families, and a manager's house. Their intentions reportedly were good, and management

sincerely believed that they would intercept the Annie Laurie vein. Like most other ventures, however, for every dollar's worth of stock sold, much of it went into the promotor's pockets.

One man claimed to have received a vision from the Lord showing him where the allusive vein could be located. The vision told him that "on the large shale mountain southwest of Kimberly, there is a line where the color changes." This was where he should dig. For three years he and his faithful followers clawed their way into the mountain; then, disillusioned, they abandoned their efforts and faded away.

In 1903, the *Mining Review* blazonly announced "that the entire Gold Mountain district is destin-

Photo courtesy Utah Historical Society

Lower Kimberly 1899

Lower Kimberly, 1899 view looking north. K&S Store first on left. Lawson Hotel, third on left. Opera House high on bank, center, right. Photo courtesy of Utah Historical Society.

ed to be the leading gold camp in Utah in the near future, there can be no doubt in the minds of any who have made even a casual investigation of the country. The Annie Laurie is already a famous mine. Through the acquisition of the Bluebird ground recently, the Annie Laurie has enough ore in sight for years to come."

Mining patent laws required that each claim holder must do at least $100 worth of work each year on his claim to hold it. Each year, by the season's end, many had given up in despair and left the mountain or moved to new claim. On January 1st, these abandoned claims were again up for "grabs." Each year on December 31st, a stream of prospectors trekked to the mountain and watched over the area they wished to claim. Then at the stroke of midnight, they marked the new claim and the rush was on to the courthouse in Junction. Skis and snowshoes were a must in most cases. Cunning, strength and endurance won the new claim.

CHAPTER 4

Kimberly and Its People

In Kimberly, Number Four Annie Laurie became the center of the mining operation and the nucleus of Upper Kimberly. The tunnel reached the innermost heart of the mountain. A network of tunnels, stopes and rises fanned out in all directions chasing the sometimes massive and often allusive ore veins. Eventually many of these offshoot tunnels linked with other diggings; Annie Laurie Number One, Two, Three, the Hastings and the Bluebird network. These latter tunnels entered the mountain approximately one mile south and several hundred feet higher than Annie Laurie Number Four.

The complex at the entrance of Number Four Annie Laurie became a beehive of activity. Long snow sheds stretched to the loading area for the tramway, an innovation that carried ore one-half mile down the hill to the huge mill in Lower Kimberly. Feed storage sheds, mule stables, mining supply rooms and an office joined the blacksmith shop and sawmill complex. This Upper Kimberly compelx was the hub of the mining activity just as Lower Kimberly became the center of the milling operation.

From the beginning, the Bluebird Mine located a few hundred yards east, southeast of the original Number One Annie Laurie, fired great expectations. Eager picks and shovels clawed three tunnels through solid rock chasing the traces of hopeful signs that might bring prosperity. Only Number Three Bluebird yielded the mountain's treasures in any substantial amounts. Located at the 8,900 foot level, it eventually became an important segment of the Annie Laurie system.

The Hastings Mine, located halfway between the Bluebird and Number Four Annie Laurie, also produced substantial "pay dirt." Begun by Jack Hastings, it, too, became part of the Annie Laurie underground complex.

Inside the network of tunnels, rises and stopes that inched their way throughout the mountain, wheelbarrows and ore cars on steel rails hauled ore to chutes to be gravity fed to waiting ore cars in Number Four Tunnel. A train of ore cars pulled by a mule then began the long trek out of the tunnel. The sure-footed mule picked its way across rail and tie by the light of a flickering carbide lamp mounted on its forehead.

Some Kimberly stories tell of mules going blind in the mines from never seeing the light of day. Such stories were fabrications. The mules were stabled in a corral area of the snowshed complex, and in actuality, saw more of the daytime light than the miners who remained in the mines all day. The mules made several trips out of the mine per day.

Huge ore wagons pulled by four- and six-team horses hauled the ore from the Bluebird tunnel down the steep mountain road to Number Four Annie Laurie, the beginning site of the tramway which carried the ore to the mill in Lower Kimberly. The tramway consisted of a trestle with two sets of tracks to accommodate the gravity operated ore cars. These cars were pulled by cables; the fully-loaded car at Number Four tunnel providing the gravity power to pull an empty car from the mill in Lower Kimberly by a system of cables and pullies. This modernization sped the ore from the central collection point considerably. Frequent breakdowns plagued the tramway, however, and these maintenance costs added to the now soaring Annie Laurie expenses.

Work in the new Number Five Annie Laurie near the mill in Lower Kimberly progressed at a frustrating pace. Great hopes hung on its development, however. In 1903, the management boasted that upon reaching theAnnie Laurie vein, the capacity of the mill would be doubled. Mudholes caused by ancient glacial action similar to those in Sevier added to tunnel maintenance. And like the Holland in Sevier, cave-ins were common with the shifting earth crushing the timber like matchsticks. Except for mudholes, most of the tunneling was through barren, solid rock.

Labor problems arose hampering progress. In 1903 Sidney Black reported, "The organization of the Western Federation of Mines is in a flourishing condition in Kimberly." The union claimed 75 members in good standing. Meetings were held every Friday night in the union hall in Upper Kimberly.

Workers continued to flock to Kimberly from far and wide. Many stayed only long enough to finance the next leg of their fortune-seeking journey. Some settled into the community life and remained on the mountain until the collapse of 1908. Some remained another 30 years searching, sweating and clawing for the one big discovery that all men dream of.

Among the colorful latter group were several Englishmen known as "Cousin Jacks." Born to the mines, most started working in the tin mines in Cornwall, England, at the age of 11. This was customary. Upon reaching the age of 18, the lucky ones went to Portugal as mine foremen. After several years in Portugal, they set a course for America and the West. These men were described as exceptionally strong and usually good natured. One of them by the name of Jim, however, proved to be an exception. He became a shift boss in the Annie Laurie. When drunk, he was known as "the meanest little man on the mountain." He reportedly fought anything that moved including his wife. This indiscretion proved to be painful one night, however. He came home during a drunken tirade and tried to get into the house. His wife locked

the door, and he continued to batter it. The wife calmly took a tea kettle of hot water, slipped out the back door, climbed onto the roof, and drenched the staggering slob in hot water. "It hurts worse than a thrashing," he admitted, telling about it later.

More than once, this man's malicious moods caused him pain. One day as he walked passed a mule in the tunnel he burned the mule's rump with a carbide light. "You'll pay for that, Jim," Fred Tilton warned him. A few days later, Jim again had to slip past the mule in the narrow tunnel. As he reached a point between the mule's belly and a large post, the mule leaned against him breaking two ribs.

With the creation of the community on Gold Mountain, the birth of new citizens came naturally. The first baby born in this bustling camp arrived on February 15, 1901, Mr. and Mrs. Mike Fitzgerald being the proud parents. Fitzgerald owned the Magnolia group of claims on the mountain.

Undoubtedly, the most famous of the new babies born in Kimberly was Ivy Baker, born to Mr. and Mrs. Orange Baker in 1905. Miss Baker later became known as Ivy Baker Priest and served as United States Treasurer under the Eisenhower Administration. Her signature appeared on all U.S. currency totaling $62,800,000,000 during this period. In 1966 she was swept into office of California State Treasurer and served for two terms. She was the first woman constitutional officer in California history.

During Kimberly's short life, three known doctors rendered their services to the community. Dr. Foute's stay was relatively short and little is known about him. Less yet is known of Dr. Kinerbye who is recorded as living in Kimberly in 1903. Dr. Steiner's name was more prevailing. His home adjoined his office. Mrs. Josephine Pace described the house as she remembered it:

"On the outside, the house looked just like the rest of the town's lumber shacks, but inside, it was pure elegance. The chairs in the parlor were covered with red velvet, and the carpet was covered with flowers. In full view, the stairs leading to the bedroom above were covered with red carpeting. There was a small table near the door that held a silver dish for calling cards, and the house had a really, truly bathroom with a long tin tub, and you could pull the plug and the water ran out."

IVY BAKER PRIEST
TREASURER

STATE OF CALIFORNIA
OFFICE OF THE

Treasurer

SACRAMENTO 95809

July 3, 1973

Mr. Dean F. Herring
1575 Willowmont Avenue
San Jose, California   95118

Dear Mr. Herring:

I was most interested to learn that
you are writing a history of Kimberly, Utah.
I was born there. My father worked at the
Annie Laurie Mine, I remember hearing my
folks say, but we left when I was very young
and moved to Richfield, and later on up into
Summit County. I have never been back since.

I believe that you would know more about
the history of Kimberly than I, and I would
very much like to know more about the town
myself. Anything you can tell me would be
much appreciated.

Arthur Baker is a new name to me;
however, Steve Baker was my uncle.

I'm sorry I can't be of more assistance,
but I do send you all good wishes for success
in all of your endeavors.

Sincerely,

Ivy Baker Priest

Ivy Baker Priest

The doctor's office was well stocked with remedies and equipment. The floors, scrubbed twice daily, were bleached white. In this boisterous community of up to 1,200 people, the doctor was never at a loss for patients. New babies, accidents, gunshot wounds and chicken pox barely scratched the surface. Josephine Pace tells of her father, Charles Skougaard, helping Dr. Steiner amputate a miner's leg in the mine in order to extract him from under a rock which had crushed the leg. Skougaard had to hold the miner during the ordeal. "Dr. Steiner gave my father some pills that night after the ordeal, but even then, he wasn't able to sleep," she said.

Under lax safety rules, accidents were the rule rather than the exception. The following are but a few.

While assisting an oiler in the mill, the jacket of Mr. Charles Hepler, an electrician, was caught by a projecting piece of shafting. He was thrown around the shaft and his foot caught in the signal rope which cut off the machinery. He escaped with a few bruises. Another man was electrocuted when his head touched a trolley wire. Joe Genetti was killed while sawing a timber in the mine when the mine caved in on him. Ernest Paulsen caught his jumper sleeve in a gear. His arm was badly mangled. Floyd Weed, superintendent of the cyanide department, plunged head first into a cyanide tank when a board broke while he as making repairs around the tank. He was rescued from the five-foot deep tank and although he had swallowed some of the solution, he was given antidotes and survived.

Primitive medical attention of the times becomes graphic in the description of a blasting accident:

"May 28, 1903. Dennis Murphy met with a terrible accident in the Annie Laurie Mine Thursday evening which will probably cost him the sight of both eyes. As he started the six o'clock shift, he was sent into a stope in which no work had been done in several months. About ten minutes later, an explosion was heard. Fellow workers rushed to the stope, and found him unconscious, his face badly bleeding. He was taken from the mine, but did not regain consciousness for one-and-a-half hours. When found, one eyeball was completely out of its socket, and was hanging on his cheek. The other eye was badly injured. He was sent to the hospital Friday. It is evident that an unreported missed shot in the face of the stope exploded when Murphy's pick struck it. Mr. Murphy had been at the camp a year. He has a wife and two small children who are in very poor circumstances. He is 45 years old."

In 1906 five men were trapped in Number Four tunnel for one-and-a-half days when the tunnel caved in. They were rescued with no injuries.

One story tells of a cave-in when two men were trapped deep inside the mountain. Realizing that their rescue would take considerable time, they carefully rationed their lunches and candle supply. They divided the lunches into seven-day portions and braced for the long wait. They had no watches. The seven-day food supply ran out and bravado gave way to apprehension. Then rescue came. The men crawled out of their near tomb. "What day is it?" was their first question.

"You've been in there a little over eight hours," they were told.

Families on the mountain brought social life common to all societies. Six months out of the year, skiing and coasting parties attracted the younger generation. Toboggan runs from Upper to Lower Kimberly attracted young and old alike. Snow banks remained in some ravines all summer affording year-round playgrounds. Dish pans and barrel staves served as sleds and skis. On summer nights, boys and girls rode horses over the moonlit mountain trails.

Parties and night life in a mining community leaves memories that words can't describe and minds can't forget. Kimberly was a textbook on mining town life. Saloon doors, open to the streets, revealed busy card tables and boisterous laughter echoed from within. One evening an earthquake shook the complacency of this idyllic setting. Men bolted from the saloon with cards still in their hands and with money left laying on the tables. One story tells of a man called "Stuttering Ohlie" during the earthquake. He was blocked at the door by a large dog. He kicked at the dog. "G-g-git the h-h-hell out of the w-w-way. If you c-c-can't run, let somebody r-r-run that c-c-can."

Mrs. Josephine Pace describes a night at the camp at Fish Creek when a young man read a script of a play that he had written. "The group built a big bon fire and everyone sat around in a circle while Erastus Bean read his play aloud. I was breathless even though I didn't know what any of it meant. The play made its author famous

Poised high on a hill, the lodge overlooked lower Kimberly. A social center in the early days of Kimberly, it was the scene of elegant parties. Photo about 1900, courtesy of Ed Tilton.

The band on the Fourth of July marching through Upper Kimberly. Photo courtesy of Utah Historical Society.

throughout the western Mormon territory."

Much social life centered around friends and neighbors. Young ladies gathered and spent evenings gossiping and embroidering. On Sunday, a community church meeting gathered at the schoolhouse. Reverend John Matier, a Presbyterian, or sometimes a Mormon Elder, delivered a sermon. A story tells of the Reverend's attempt to reform a young lady of dubious reputation. One

Sunday, while greeting his parish after the sermon, the young lady passed by without stopping. The Reverend called to her. "Helen, I prayed a solid hour for you last night," he said.

Helen turned and smiled. "Shucks, Reverend, if you had have sent word, I'd have been right over."

Elaborate social functions centered at the lodge. Poised on a hill overlooking Lower Kimberly, its

Kimberly businessmen, 1902. Left to right: Russell Ivy, barber, Frank Latchow, butcher, Clair Weber, bartender and popular baseball player. Photo courtesy of Charles Hansen.

Lambertson shoe shop, last of the buildings to disappear from Lower Kimberly business section. Photo by Dean F. Herring, 1954.

mansion-like windows exposed the teaming activity below. Aside from the huge mill, the lodge was the most imposing building in Kimberly. The structure began as a log cabin housing a Pelton wheel which turned an electric generator. It grew to a floor space of 4,400 square feet. It boasted of a 20 by 27 foot sitting room with portiered bay windows and cushion seats, five bedrooms, two baths, a large kitchen, a dining room, a large cellar pantry, and a woodshed. Although it was not a boarding house, engineers and surveyors were usually invited to stay there.

An article in the *Richfield Reaper* in 1903 records a party held there:

"Misses Mary and Ida Gould entertained in a most delightful manner, a number of friends at the lodge at Kimberly on Monday night. The entire building was brilliantly illuminated with electric lights. Music, games, and guessing contests were indulged in until midnight. Prizes were won and dainty refreshments served in the dining room which was tastefully decorated with evergreens and American flags."

Mrs. Josephine Pace described her childhood memories of the lodge: "The parties at the lodge were very elegant. From my vantage point belly boost at the top of the stairs, I could see it all. The ladies in their dresses with demitrains which they would hold just high enough to show the lovely petticoats underneath them. And mother, of course, was the prettiest of them all with her dress of brown voile, a rustling stiff petticoat of green taffeta, and her locket on a chain. One of the gentlemen passed a tray of tiny glasses filled with wine which the ladies sipped slowly as I would eat my Sunday ice cream. The tables also had a deck of cards, and a game of 'Whist' was soon underway."

As in most frontier towns, drinking and gambling were common forms of entertainment. A beehive of activity surrounded the saloons. The tenth of the month was payday and gamblers, commonly referred to as "tin horns," converged on Kimberly and quickly separated the weak from his paycheck. It wasn't unheard of for a winner at blackjack at a saloon to lose it all to a blackjack on the dark trail home. At least one man was killed while going home to his cabin after a night of drinking and winning at gambling. He was reportedly struck over the head at the road fork near Surprise.

On the Fourth of July, a parade wound its way down the road through Upper Kimberly, then down through the main street of Lower Kimberly. Children with boiler lids, dish pans, whistles—anything to coax a noise—brought up the rear in a bedlam of commotion. In the center of town a greased pole was set up in the middle of a horse-watering trough. The prize was $100 for the first man to climb the pole.

Rock-drilling contests with drill steel and sledge hammers brought the winner $50 and competition was keen. Tugs of war, races and pie eating contests filled out a busy day.

In the evening, the lodge hosted a children's dance. One Fourth of July, a prize fight was held between Jimmy Burns and Joe Wadinski in the spectator-packed opera house.

At home, families gathered and members took turns cranking a home-made ice cream freezer. Snow was readily available throughout most of the summer for this popular bit of luxury.

The opera house sat high on the east side of the main street in Kimberly. A flight of stairs led up to the front door. This center of entertainment boasted of a stage 20 feet in depth and a mechanically operated curtain. The dance floor of two-inch maple boards measured 30 by 60 feet. More than 80 chairs ringed the walls for those with tired feet. Traveling entertainers made Kimberly part of their circuit.

Throughout the summer, Saturday night dances at the opera house became an institution, drawing participants from the valley as well as from the far-flung reaches of the mountain. An orchestra, centered around the Joe Nelson family, was popular not only in Kimberly, but throughout the state. Couples perspiring from the racing tempo of the dance floor, strolled off along the mountain trails to cool off and find seclusion.

A baseball park was cleared off below Lower Kimberly and this sport became part of Kimberly life. At this 8,000-foot elevation, bases were run slower but with no less enthusiasm. James Sorensen was a popular pitcher. "Just wait until I get back in shape, and I'll show them a thing or two," he said as he returned to Kimberly from a winter in the valley. Clair Weber's name was legend in the sport.

Children's games included one called "Livery Stable." The willowy and profuse quaking aspen were bent toward the ground and used for horses. The Keeper of the livery stable would "rent them out."

At Christmas time, the lodge was the popular spot on the mountain. Chains of popcorn, cranberries, and paper decorated the rooms and the candle-lit Christmas tree. Presents were exchanged by the cheery crackling fireplace. The laughter and song inside the frosted windows contrasted sharply with the deep glistening snow outside.

Winter storms, though accepted as a matter of course, at times raised havoc among the dug-in residents. A 1902 news article in the *Richfield Reaper* cites such a storm:

"On Gold Mountain, there has been, all told, about five feet of snow, and it drifted as fast as it fell. The continued wind which has prevailed there during the entire storm, has made it very difficult to get up and down the mountain. Sunday morning, the wind blew the frozen snow with a velocity so fierce as to break a great number of windows of the buildings. In Kimberly, Snyder's and Keeler's saloon had hardly a whole pane left in the structure. The lateness of the fall, and the fact that it has never been frozen into masses, makes snowslide to be feared. Some parts of Kimberly are dangerously located, and will be damaged if a slip of snow comes down the hillside above. One slip occurred at the Bluebird Mine Saturday or Sunday covering up the mouth of the tunnel. No one was injured. It is not known yet to what extent the buildings and other equipment were damaged."

Over the years, stories of Kimberly's red-light districts have surfaced and been challenged. Most, it seems, have been stretched out of proportion. The best authorities insisted there was only one house of ill repute ever operated on the mountain. This was a four-room house located above Upper Kimberly and east of the Hastings Mine near a meadow. Girls were imported from Price, Utah, an important coal mining area. Lack of business and pressure from housewives reportedly forced early closure. Undoubtedly, the free-lance prostitute operating from barmaid or waitress jobs plied their trade on the male-dominated mountain, but facts about them have been lost to time.

Stories of the notorious Kimberly jail also have become legend with the passage of time. Most of the notoriety, no doubt, grew from its prominent location on the bluff at the north end of town. Through the years, curiosity has sparked questions about the small brick building. Questions drew answers, not always correct, and re-telling distorted the stories. Some stories paint a picture of jail cells filled to capacity with murders, robbers and drunks.

Mr. Fred Tilton, one time part-owner of the Annie Laurie Gold Mines, tempered this picture. From the time of his arrival in Kimberly in 1905, until the bankruptcy of 1908, he stated that he seldom knew of anyone being in the jail. That there was crime in Kimberly, there is no question, and the two small iron-gridwork cells, no doubt, held their share of interesting occupants. A pot-bellied stove, almost the only furnishing in the building, was ample for heating. The four grid-iron, wall-mounted folding bunks, though hard, were sturdy and functional. The jail was claimed to be the strongest in the state of Utah.

Some insight to the occupants that served time there is gained from these articles from the *Richfield Reaper*:

(1903) "For a few minutes, there was excitement at Kimberly last Saturday afternoon about two o'clock. The successive crackling of pistol fire caused a rush of people to the streets. In the front of the Kimberly Hotel, two men were seen emptying their guns at one another. H. Von Martin succeeded in getting between the two combatants and separating them. Neither was hurt. They had been imbibing rather freely of the flowing bowl, and

The young crowd at Kimberly, 1905. Front row (L to R): Ron Seegmiller, Ben Carter, Otto Fransworth. Rear row: Joe Jensen, Arther Baker, George Hansen, Dwight Metier, June Seegmiller, Jess Bean. Ben Carter was involved in two shootings at Kimberly. In the most publicized, he shot Lawrence Hamel in the cheek, the bullet stopping in Hamel's teeth. Hamel spit out the bullet and beat Carter to near unconsciousness.

The Kimberly jail cell, stripped of its brickwork shell, 1954.

got in a dispute over a claim. The two principals, R. J. Gibson and J. Jacobs, are well-known throughout Utah, being old-time prospectors."

(1903) "A quarrel with a serious outcome is reported from Kimberly. For some days, Harry Roberts had been deeply incensed at an insult that was said to have been offered his wife by one of the miners, name unknown. Sunday evening, they met. The miner denied the charge and the matter worsened. Roberts, feeling not justified in taking more under the circumstances, knocked his antagonist down and gave him a number of kicks. One struck the man in the eye, and is thought to have put it out entirely."

Also in 1903, a Richard Urideen was stabbed in the neck and abdomen during an argument in Upper Kimberly.

As in most neighborhoods, petty crime and hooliganism plagued the populace. Stories tell of nightshift workers, trying to sleep in tents, becoming targets of young pranksters throwing rocks and cutting tent ropes, allowing the tent to collapse on them. Barking dogs mysteriously died of poisoning, and outhouses rolled down hillsides in the night.

Some evidence indicated that investigations and justice were based on emotions. A well-known miner who worked at the Silver Queen east of Kimberly occasionally walked to Kimberly to find entertainment. Stories spread that he found it with the wife of a Kimberly resident. Late one fall even-

ing after a rendezvous with his lady friend, he left Kimberly for his cabin at the Silver Queen. He never arrived. Heavy snows prevented a thorough search for him. In the spring, a search party set out to look for him. One member of the party was the aggrieved husband, and it was he who seemed to know where to look for the lost miner. When the body was found, it was clad in a long heavy mackinaw. The belt was still secure and snug around the waist. He had been shot. His death was ruled accidental, self inflicted.

The most publicized occupant of the Kimberly jail was undoubtedly Benjamin Carter. On May 20, 1908, during a drunken argument, Carter drew his gun and shot a man by the name of Lawrence Hamel. The bullet struck Hamel in the cheek knocking out several teeth, but he failed to go down. He spat out the bullet and shattered teeth, then proceeded to beat Carter to near unconsciousness. Carter was sentenced to ten years in the state penitentiary.

A frequent visitor to Kimberly was another well-known outlaw, LeRoy Parker, or later known as Butch Cassidy. Born in Circleville, Utah, 40 miles south of Kimberly, his visits to Kimberly were peaceful. To Charles Skougaard, who sometimes carried the mail and payroll to Kimberly, "LeRoy was just a boy from Circleville who joined a wild bunch of cattle rustlers." He was respected as a man who never broke his word nor betrayed a friend. One member of the Cassidy gang married

William (Billy) Morrison's stage leaving Monroe for Kimberly.

a sister-in-law of Dr. Stiener, the Kimberly doctor. Stories have suggested that Cassidy did, indeed, plan a robbery at Kimberly. The fact that he was known and respected there seems to rule out this possibility.

The stagecoach forged the link between Kimberly and the outside world. Orson P. Washburn ran a "first class" stagecoach between Kimberly and Monroe twice a week, on Tuesdays and Thursdays. The fare up the mountain was $1.25. Down, it was $1 and round trip was $2. Accommodations were described as excellent. He also delivered "first class" green vegetables and fruit.

The William Morrison stage, headquartered in Monroe, ran each Tuesday and Friday. The noonday meal at the Robinson ranch in Clear Creek Canyon was a welcome break for the weary passengers.

The Oluf Olsen stage made the trip from the railroad station in Vaca to Kimberly in the morning and back in the afternoon. He carried the mail and passengers.

Other stages during this period included one operated by Burt and Frank Shepard.

The grueling trip up the rough mountain roads sapped the strength of lather-drenched horses and jolted the aching bodies of the tired passengers. Rest stops were frequent, and none was more refreshing than the one at the "Water Cress." There, a spring-fed stream crossed a small meadow in a grove of quaking aspen trees. Lush clumps of water cress covered the banks of the stream. The horses and passengers drank, then while the horses caught their breath, the passengers stretched their legs and nibbled on the succulent water cress.

It was here at the Water Cress that another enterprising businessman by the name of Jacob Barney built a slaughter house. Bones of the butchered animals can still be found on the hillside below this site.

Arrival of the stage was the big event of the day in Kimberly. With the stage came letters from

Jacob Barney, an enterprising businessman, ran a slaughter house at the Watercress one mile below Kimberly.

The Kimberly band in Lower Kimberly, 1905. View looking south, Keeler's Saloon nearest band. Lawson Hotel, building #4, K & S Store, last building, center of photo. Photo courtesy of Utah Historical Society.

Lower Kimberly business section same as above. Photo courtesy of Utah Historical Society.

loved ones and packages from far-away places. Friends and relatives returning home and interesting strangers from the outside world added to the excitement. Salesmen with beautiful serge material, table cloths and dresses brought the outside world nearer to the mountain. The men from the east with stylish straw hats, good manners, and accents attracted the ladies. The event is best described by Josephine Pace:

"We were never late for its arrival. Four, and sometimes, six horses covered with lather arrived in a swirl of dust. Little Billy Morrison of Monroe would pull on the reins bringing the horses up sharp as he stopped to unload mail from the stage. Then on the stage traveled, down the draw, and across the creek to the Lawson boarding house to let passengers off."

The stage ran the year around. During the summer months, extra coaches were used to handle the increased passengers and mail. Winter snow slowed but did not stop the stages. Little attempt was made to keep the roads open to wheel traffic. The four- to six-foot snow pack presented a formidable problem. Removing the wheels and installing runners proved to be the most practical solution. The changeover was made at various locations along the route depending upon the snow depth. At times, the entire trip was made on runners.

At times in the deep snow, only the bushes alongside the roadway defined its path. The lead horses, in the words of Oluf Olsen, "they were used to it. They'd just jump up and down and break the road, and I'd make the 'wheelers' pull the load." Mr. Olsen had two four-horse teams which he rotated each day.

Travel on the Kimberly stages sparked many stories. Ninety-two-year-old Oluf Olsen glowed as he told of a speed run hauling an official down from Kimberly to Vaca to catch the train. "I yelled, 'Hang on,' " he chuckled, "and we went through the creek at a gallop." According to Olsen, his wet and disheveled passenger both cursed and complimented him for the successful ride, and gave him a five-dollar tip.

Drunken passengers added to the stage driver's woes. Olsen said he occasionally was forced to tie them to their seat for their safety. He chuckled again when he told of a mishap in the snow:

"The road was narrow and dangerous. I was bringin' Ohlie Nielson, a gambler, they called him 'stutterin' Ohlie.' I was bringin' him and a newly-wed couple up from the valley on a sled. I asked Ohlie if he would get out on the 'uphill' runner when we went around this steep turn to keep it from turnin' over. Ohlie said, 'Let 'er tip. I'd kinda like to see what's goin' on under that blanket.' "

According to Olsen, no one stood on the uphill runner, and the sled tipped over, dumping the newlyweds into the snow. "I never saw such a thrashin' of legs as they tried to untangle themselves," he laughed. "Old Ohlie, he stuttered, 'I d-d-don't know w-w-what it d-d-did for them, but it s-s-sure was a s-s-show for us!' "

Stage driving had its serious moments, according to Olsen. "I was comin' up one night all alone. I had a headlight on the hack, ya' know. And I got to a certain point on the shale dugway. There's a big sharp turn there, and these two men were comin' down. They had these big lights on their heads. When they got on that point, they turned their lights out. And that night, I had all the money for each store. Six bags of gold and silver under my seat in a little suitcase. And a lot of 'em got on to that, ya' know. Come up before payday. These guys got on to that. It would have been the best pickin' they'd ever had. I got up to that turn and my light shown on 'em. One stood on the upper side, and one on the lower to grab my horses. And when I seen 'em, I gave a yell at my horses, and whistled at 'em, and they broke and ran right through 'em. They grabbed for the leaders, ya' know, to catch 'em, and they missed the leaders. One of them grabbed a tug on one of the wheelers and it pulled him right under the rig. And the other'n got the other horse, or tug, or somethin', and he got throwed over the edge. And I never stopped to see if they were killed nor nothin' else."

Another serious moment came when a stagecoach with two armed guards carrying a gold brick to the valley lost the brick which apparently merely bounced out. An honest man, John Waters of Monroe, relieved the resulting trauma by turning the brick in after he found it in the road.

Large freight wagons pulled by four- and six-horse teams plied the mountain roads. From the valley, they carried produce—hay, oats, butter, eggs, vegetables, fruit, machinery, parts, and building supplies—all things that a town must have to survive. The valley, in turn, prospered.

Snow travel was treacherous and grueling. Heated rocks and bricks wrapped in blankets protected driver's and passengers' feet against the cold. Heavy traffic kept the snow packed but, at times, the effort was almost too great. Wagons and bobsleds operated on a limited scale until the going became nearly impossible. Then pack trains substituted. Every effort was made to lay in winter supplies before the snow came.

Horses were the machinery that kept Gold Mountain going. Like trucks and automobiles of today, sweating horses plied the mountain trails and roads carrying travelers, passengers, supplies, mail, logs, and ore. Livery stable business was big business. Nothing was more precious to the freighter than his horses. One was once asked, "What would you do if you had a million dollars?"

"I'd buy the best damn four-horse team on the road," he answered.

Will Thurston ran one of three livery stables in Kimberly. He also kept saddle horses for rent. Mortin Christensen and Charles Leavitt operated a dairy in Lower Kimberly. The milk was carried from house to house in a large ten-gallon milk can. Each sale was dipped from the can and measured into a container. At first, the cows were taken down to the valley during the winter months. Later, hay and grain were hauled up to the mountain and the dairy ran the year around. Many households had their own milk cow. The cows ran free on the mountain coming home at night for their treat of hay or grain.

For peddlers from the valley, Kimberly was truly a "gold mine." Customers with ready cash bought the door-to-door fruits, eggs, butter, and even shoes.

For most prospectors, miners and workmen scattered at the remote outposts and claims throughout the mountain area, contact with the outside world was minimal. Travel was difficult. Even walking at this high altitude required great effort. Mail was the only contact for many, and the post office at Kimberly for some was a one-day hike. In 1900, Woodruff Sylvester, a 13-year-old boy, got the job of delivering mail to these lonely outposts. Each day, he picked up the mail at Lower Kimberly and covered the 18-mile route by horseback. The route led him to Sevier, down into the depths of Fish Creek, then south up the pipeline trail to Signal Peak, and back to Sevier where he lived with an uncle. His clientele varied between 30 and 40 customers who each paid him one dollar per month.

A small two-room structure served as a schoolhouse in Kimberly. One room built of logs and one of frame construction sat across the creek from the Lawson boarding house. Most classes were held in the spring, summer and fall. The deep snows of winter made this season vacation time. There was no grade structure, the five to 11-year-old pupils all working together. The homemade desks accommodated two pupils each. The rooms were heated by a wood-burning stove. During the cold weather, those sitting near the stove suffered from the heat while those in the back of the room had trouble keeping warm.

A school record book retrieved by Ken Lawson of Kanosh, Utah, from the debris left at Kimberly in the late forties, reflects that in 1903, teachers E. Fitzgerald and Myrlie Bartlett had 65 pupils ranging in age from five to 11 enrolled in the school. Other teachers reported to have taught there in later years were a Miss Butler and a Miss Paxton (no first name available). After the bankruptcy in 1908, a Miss Shipley and a Miss Baker taught at the school for those few remaining.

The available school records for 1903 are as follows:

Kimberly School Records—1903

Lucile Avery, pupil, age 5
  Parent, Thomas Avery
Myrlie Bartlett, teacher
Edward Bell, pupil, age 6
  Parent, Ina Bell
Delora Bridges, pupil, age 6
  Parent, M. Bridges
Emory Bridges, pupil, age 8
  Parent, Wm. Bridges
Dell Carter, pupil, age 6
Devier Carter, pupil, age 7
  Parent, A. Carter
Albert Dimick, pupil, age 6
  Parent, Albert Dimick
Lavina Dimick, pupil, age 9
  Parent, George Dimick
Elva Durfey, pupil, age 9
Ernest Durfey, pupil, age 6
Monila Durfey, pupil, age 5
  Parent, E.L. Durfey
Bryan Farragh, pupil, age 6
Katie Farragh, pupil, age 7
Robert Farragh, pupil, age 11

Kimberly schoolhouse, foreground, view looking north. Stable, center left, K & S Store, upper left. Dr. Steiner's home and office, extreme right near center. Opera house, high building upper right. Photo courtesy of Charles Hansen.

Harry Farragh, pupil, age 9
  Parent, James Farragh
Pearl Finley, pupil, age 6
  Parent, C.E. Finley
E. Fitz Gerald, teacher
Angelia Gibson, pupil, age 7
  Parent, R. Gibson
Elsig Hales, pupil, age 8
  Parent, W. Hales
Clarence Hyland, pupil, age 8
John Hyland, pupil, age 6
Lizzie Hyland, pupil, age 10
  Parent, John Hyland
James Hyland, pupil, age 6
  Parent, Patrick Hyland
Hope Jensen, pupil, age 6
  Parent, Mrs. P. Christensen
Ruth Johnson, pupil, age 7
  Parent, A. Johnson
Hilda Johnson, pupil, age 7
  Parent, Archie Pinney
Ivan Keeler, pupil, age 7
  Parent, Arthur Keeler
Ester Kinerbye, pupil, age 8
Gerda Kinerbye, pupil, age 7
  Parent, Dr. Kinerbye
Johanna Lambertson, pupil, age 10
Laura Lambertson, pupil, age 8
Raymond Lambertson, pupil, age 6
  Parent, M.L. Lambertson
William Lawson, pupil, age 7

  Parent, Andrew Lawson
Vera Ling, pupil, age 8
  Parent, William Ling
Lara Merritt, pupil, age 10
Martha Merritt, pupil, age 6
Martin Merritt, pupil, age 6
  Parent, Martin Merritt
Melvin Mitchell, pupil, age 7
  Parent, Wm. Noslain
Delia Nay, pupil, age 10
Eldred Nay, pupil, age 6
Foster Nay, pupil, age 8
  Parent, E. Nay
Ethel Parkinson, pupil, age 8
  Parent, W. Hales
Edith Pinney, pupil, age 9
Nora Pinney, pupil, age 6
  Parent, Archy Pinney
Edward Pitts, pupil, age 7
Mary Pitts, pupil, age 5
  Parent, James Pitts
Lennard Pitts, pupil, age 7
Vivian Pitts, pupil, age 9
  Parent, Thomas Pitts
Goldie Pyle, pupil, age 9
  Parent, Henry Pyle
Roland Robinson, pupil, age 6
  Parent, Walter Robinson
Beatrice Robinson, pupil, age 8
  Parent, W. Robinson
Jimmie Robinson, pupil, age 7

Levell Robinson, pupil, age 5
  Parent, Ina Robinson
Florence Shelton, pupil, age 7
  Parent, William Shelton
Winneford Smith, teacher
Bryan Sorensen, pupil, age 6
  Parent, James Sorensen
Thomas Stanforth, pupil, age 5
William Stanforth, pupil, age 7
  Parent, William Stanforth
Grace Taylor, pupil, age 6
  Parent, George Taylor
Vagus Thompson, pupil, age 6
  Parent, J. Thompson
Bryan Thurston, pupil, age 6
  Parent, William Thurston
Anna Wilson, pupil, age 6
  Parent, H. Wilson
Francis Winn, pupil, age 6
Lucile Winn, pupil, age 9
  Parent, Frank Winn

Cummings boarding house in Upper Kimberly, 1930. Photo by Floyd Herring.

Housing developments covered the hillsides in and around Kimberly ranging from drifters' tents, shacks and cabins to well-built company-owned houses. There were approximately 20 of the latter in Lower Kimberly and approximately the same number in Upper Kimberly. The company also owned the lodge about a quarter of a mile above Lower Kimberly. The company furnished free lumber to employees wanting to build their own shelters. Since ownership was only as permanent as the job, most of these employee-built houses reflected impermanence and cluttered the mountainside randomly and in clusters.

Most company-owned houses in Kimberly were of near identical construction—square with a pointed roof, and divided into four almost-identical rooms. Only the kitchen was different, furnished with a sink and running water. The walls of solid pine boards were covered with a cotton muslin called "factory." Wallpaper was pasted onto the cloth. The near identical furnishings, also company owned, included a small wood cook stove, a few shelves nailed to the walls to serve as cabinets in the kitchen, a plain wooden table and four to six chairs, a small wood-heating stove in the living room, and iron bedsteads and springs in each of the two bedrooms. They rented for $5 per month. The managers' houses were different. One located in Lower Kimberly and one in Upper Kimberly were larger and were two of the four houses in Kimberly to boast of a bathroom.

In Lower Kimberly, the bulk of company-owned houses clustered near the business district on a meadow around the Lawson boarding house, along the road leading out of town and near the mill. Away from the main clusters, many houses, tents, cabins and shacks set in groups or scattered by themselves away from the settlement.

In Upper Kimberly, the main body of housing clustered near Number Four Annie Laurie tunnel, then spread south scattered through the trees up the hill toward the Bluebird Mine. The Cummings boarding house, a large log two-story building owned and operated by Mr. and Mrs. William Cummings, set by itself a few hundred yards up the hill from the main settlement, and a stone's throw from a delightful meadow. A blacksmith shop and two stables set just across the meadow from the boarding house. A settlement of Italian workers lived in self-constructed shacks and tents beyond the boarding house scattered among the heavy stands of quaking aspen. Nails driven into the trees supported pots and pans. Five- and ten-gallon cans served as stoves for cooking and for heat. Pine boughs sufficed as mattresses.

Wood was the only fuel available on the mountain and was available in abundance—at first. As the town grew, so did the distance to the fuel sup-

Lower Kimberly, 1928. Lawson boarding house in foreground. Lodge, upper left. Business section, lower right. Photo by Ed Tilton.

plies. The 80 cords of wood per day required by the huge mill depleted the hillsides. Bert Shepp constructed a chute on the east mountain to speed wood recovery from its slopes. His head was torn off when a piece of cordwood jumped from the chute killing him.

Most houses had attached lean-to woodsheds for wood storage. The sound of hand saws and axes cutting wood echoed across the canyons each night in the still mountain air.

A small dam in Mill Creek about one-half mile above Upper Kimberly provided a settling pond for the settlement's water supply. Water was piped to most of the houses in both Lower and Upper Kimberly. The dam also provided water for the fire hose at the mill. This two-and-a-half inch diameter 250-foot man-pulled hose was virtually the only fire protection on the Kimberly side of the mountain. A similar fire hose protected the mill in Sevier.

Lawson residence, 1905. Left to right: Kate Lawson, Martha Lawson, Anna Hansen—two unknown. Lawson children on steps. Photo courtesy of Charles Hansen.

Upper Kimberly, 1928. Kimberly was a treasure chest of vacant houses. Most were just as occupants had left them in 1908. Photo by Ed Tilton.

From a few tents in 1889, Gold Mountain had mushroomed into a complex society. On the west side of Kimberly's downtown business section, Andrew Lawson operated a two-story hotel. Though described as "modern," its only inside plumbing was water to the kitchen sink, three washrooms upstairs and one downstairs.

To the south among others was the K & S (Krotky & Skougaard) store and a livery stable. Several smaller businesses thrived on down the street to the north including Hans Hansen's restaurant, Elmer Nay's blacksmith shop, Sid Black's Saloon, and Jack Willardson's Saloon and Poolhall. Keeler's Saloon also served as a bank cashing worker's checks. A livery stable and the jail set at the north end of the business district. Chris Johnson and a Mr. Larsen ran a livery stable.

On the east side of the street, Dr. Steiner's office/residence set approximately 10 feet above the road level on the hillside. Immediately to the north, the Kimberly Merc—owned by Fred T. Tilton and Max Krotki—shared the small water pipe with the doctor's office. Krotki used to complain that you "had to add whiskey to the water to get enough to drink if the doctor was using any."

John Sandberg's store, a laundry, Russell Ivy's barber shop, Lambertson's shoe shop, Frank Latchow's butcher shop, the opera house (which also served as a dance hall) and others lined the street for one-quarter mile. Among other businesses recorded in Kimberly were A. J. Moore and Magnus Nielson's butcher supply house, and Abner and Gay butcher shop.

Across the creek to the west from the main business district, the large two-story Lawson boarding house set on the edge of a small meadow. Operated by Kate Lawson, it was one of the more imposing buildings on the mountain. A giant four-oven wood-burning stove with 12 removable lids dominated the kitchen and the conversation of those visiting it. The boarding house shared the meadow with several houses. Down the creek to the north, and on to the west toward the mill, dozens of tents and shacks sprawled along a network of trails. Across the creek and a hundred yards up the hill from the boarding house, an abandoned tunnel was utilized as a cold storage facility for the town. Snow was packed into the tunnel during the winter months to provide the low temperatures needed for storage. Shelves installed on the tunnel walls were marked with sharing-residents' names. Butter, milk, eggs and other perishables filled the shelves during the summer.

By 1908, Kimberly boasted of three general stables, two hotels, three barber shops, two large boarding houses (plus several smaller ones), a schoolhouse, a doctor's office, a post office, an opera house, a dairy, a laundry, three saloons, three butcher shops, and several specialty shops.

Stables rented horses to travelers heading to Sevier or other far-flung outposts. Hamner Smith, a 12-year-old boy, made regular rounds of the mountain outposts retrieving these one-way rentals. His mother, he remembered, baked up to 30 loaves of bread each day which he sold and delivered on his trips throughout the area. A Mrs. Woolsy operated a laundry using a handcrank washing machine for ten cents per hour. Eliza Baldwin cooked for as many as 90 men at a boarding house. At its peak, upward of 500 men worked for the Annie Laurie Gold Mines alone. Twelve hundred people lived in Kimberly during the summer months.

A bustling way of life stabilized and its people grew complacent. A writer for the *Mining Review* observed, "One splendid indication of the permanency of the district is the mountain roads which radiate from Kimberly to the leading mines in the district. These roads are substantially built at a cost of thousands of dollars. A stranger is agreeably impressed with these fine thoroughfares."

Few records remain of these early workers. The following is a list of some of them in 1908.

Annie Laurie Gold Mines—1908

Batisto Aimo, Miner	$3.25/day
Ambrogirilla, Miner	$3.25/day
A.L. Anderson, Teamster	$3-$5/day
C.M. Anderson, Assay Helper	$90/month
John Andreson, Solution Man	$3/month
Ishmael Arrigoni, Driver	$3/day
Peter Arrigoni, Timberman	$3.50/day
Richard Avignone, Solution Man	$3/day
D.D. Baker, Machine Helper	$3/day
J.S. Bell, Hoist Engineer	$3.25/day
Joseph Black, Electrician	$90/month
H. Boeirline, Teamster	$3-$5/day
Peter Borla, Timberman	$3.50/day
George Boyle, Solution Man	$3/day
E.M. Bush, Chief Engineer	$125/month
Andre Capatono, Miner	$3.25/day
George Cardero, Miner	$3.25/day
Den Carter, Driver	$3/day

John Carter, Miner	$3.25/day	Sisdon Hatch, Wood Cutter	$2/cord
James Castroli, Miner	$3.25/day	John Hastings, Timberman	$3.50/day
Steve Castroli, Mucker	$3/day	Arthur Hastings, Miner	$3.25/day
Dorn Cebaria, Driver	$3/day	James Hill	No record
Felio Cervi, Miner	$3.25/day	Milton Jentry, Solution Man	$3/day
John Christoferson, Mill Man	$2.50-$4/day	James Kane, Trammer	40¢/hour
Dan Coughlan, Solution Man	$3/day	Alex Larson, Watchman	$3/day
Jerry Coughlan, Cyanide Foreman	$150/month	Ernest Lindquist, Wood Cutter	$3/day
J.F. Coursey, Blacksmith	42.5¢/hour	Joseph Lloyd, Machinist	44.5¢/hour
A.B. Devine, Shift Boss	$4/day	Heber Mathews, Solution Man	$3/day
Pat Devine	No record	George Mortinsen, Wood Cutter	$3/day
Earl Dissinger, Superintendent	$150/month	Allezzander Piccine, Solution Man	$3/day
W.H. Edwards, Mine Foreman	$4.50/day	J.M. Ross, Teamster	$3-$5/day
William Elvin, Skip Tender	$3/day	Charles Shipp, Compressor Man	$3/day
E.T. Goodwin, Mill Man	$2.50-$4.50/day	Tom Simmonson	No record
J.F. Graham, Mining Engineer	$150/month	James Sorensen, Solution Man	$3/day
Fred Hansen, Engineer Helper	$3/day	John Winn, Mill Man	$2.50-$4/day
John Henry Hatch, Wood Cutter	$3/cord	Matt Worden, Solution Man	$3/day

CHAPTER 5

Financial Collapse

In June 1905, Peter Kimberley died and the mine was taken over by a British company. The new managers, Pat and John Hyland, were paid on a tonnage basis. In order to increase the tonnage, they introduced a "shrinking" system of stoping. This involved tunneling under large deposits of ore, then causing them to cave. The system was ill-adapted to the particular type of ore in Gold Mountain. Waste deposits—unusable debris—mixed with the ore greatly reducing the value per ton. The cost of milling more than offset the increased production. Oldtimers complained that the new managers "didn't know any more about mining than a hog knows about Sunday School."

The honeycombing of the mountain continued and the floors of the stopes began to cave. The operation became even more hazardous and many stopes were lost.

In 1907, financial panic swept the nation. The *Mining Review* observed, "It has been estimated that over three billion dollars has been lost in the last six months by the speculating public in the United States. In the western states, this loss will be felt but slightly, and the shaky condition of money and the metal markets are the only potent factors to be considered. Temporary curtailment of production will surely have the desired effects on the price of metals."

In Kimberly, the gold-bearing ore dwindled. The high cost of operating the huge wood-burning mill soared primarily due to the growing distances in-

volved to the wood supplies. The cost of drying the ore alone rose to three dollars per ton. In Sevier, mismanagement and lack of available ore shook the confidence of investors.

Willard Snyder, one of the men who gave the most to the mountain, apparently sensed the end of an era. He reportedly sold his interests for $1,000,000.

In January 1908, the great Annie Laurie Gold Mine was unable to meet its payroll. Butchers, grocers, merchants and saloon keepers stretched credit to the breaking point. Company representatives rushed to New York. Frantic meetings were held. The company assured the people that new money was coming from the East—that this was a temporary thing—mismanagement in the eastern offices. In the meantime, the company would operate as usual, but rather than money, company script would be used. The script, of course, would all be redeemed when the financial problems were resolved.

Under this new policy, the workers were to be paid one-third of their pay in script, to be redeemed in 30 days, and another third to be redeemed in 60 days. Often the script was not paid when due. Turnover in help skyrocketed. Miners threatened to kill the manager, and a body guard was hired. Oluf Olsen, a stage driver, paid $50 in script for a four-room house, then working in four feet of snow, moved it down to Monroe in the valley.

Weeks and months dragged on. Families, paid

in script, had only script to spend. Script was accepted only in local businesses. Grocery and merchandise stocks dwindled and could not be restocked. Josephine Pace remembered, "Cash drawers in the K & S store were jammed with script. Some with little faith moved away. Those with hope and big investments stayed on. The cottage rentals had been paid in script, and mother said we had enough of it to pay for the whole house. We lived on porridge and hope for a few years waiting for Kimberly to come back, but I guess the vein had run out."

In the end, the cold fears were realized. On June 11, 1908, the stockholders petitioned the court to appoint a receiver. Approximately 100 employees filed a lien against the Annie Laurie for wages past due. Some claims covered the entire period from January until June.

The largest creditor of both the Annie Laurie and the Sevier Gold mines was the Salt Lake Hardware. The second largest was Fred T. Tilton, owner of the Kimberly Merc. Tilton attributed his financial loss to a large extent on his associate's "inability to say 'no' to the cashing of company paychecks."

In the final settlement, the Salt Lake Hardware became owner of 80.2 percent of both the Annie Laurie and the Sevier Consolidated Gold Mines. Fred Tilton became owner of 19.8 percent of this stock. A declaration of trust filed on June 5, 1909, reflects that Charles Lammersdorf sold his remaining interests in the Sevier to the Salt Lake Hardware and to Fred Tilton for a total of $41,385.60 and $10,117.90 respectively.

Small creditors were paid off at the rate of five cents on the dollar. For most, this was financial ruin.

The mountain had surrendered a fair ration of its treasures and seemed to say, "That is enough." Destitute families moved on in deference to the uncompromising mountain. To eat, one must have money. The exodus was swift and, by fall, only a skeleton crew and the most persevering remained.

During the period between 1903 and 1908, records indicated that the milling had failed to obtain the amount of gold extraction from the ore equal to the difference between the heads (assayed value of the ore) and tails (tailings after the gold is extracted). It was obvious that some of the gold was disappearing somewhere in the process. This difference amounted to between $800 and $1,000

per month. Years later quite conclusive proof surfaced solving this mysterious "high grading." A close relative of one of the managers sold approximately $30,000 worth of gold to an assay office. The source of gold bullion is quite easily identified by the presence and the percentage of other metals. This gold had come from Gold Mountain. Further evidence of how this theft was accomplished was uncovered in 1930. A promoter came to Kimberly with the idea of beginning a new operation of the mine. Before the new company began operation, he agreed to clean up the mill area for three-quarters of the gold in the assay office slag pile. A startling discovery was that large bodies of concentrated ore, some assaying 30 to 40 percent pure gold, had been dumped into the slag pile below the bullion room. The assumption was that the person responsible had purposely dumped this concentrate out to be hauled away at his convenience. Then due to the sudden collapse of the company or fear of being detected, he was unable to complete the theft. Several thousand dollars was recovered in the slag pile.

During its lifetime, the Sevier Mines produced a considerable amount of gold, but the veins were irregular, and many faults in the unstable ground made the veins allusive. Three crucial points, each devastating, combined to strangle and eventually cause the collapse of this once promising undertaking:

The large stockholders in the Sevier Mines spent money lavishly with the expectation of large-scale expansion. They had little or no experience in this type of management, yet they elected to run the operation themselves without seeking knowledgeable advice. Most important, however, the rich lode that had seemed so promising eluded them in the unstable, fault-ridden ground. The 75-ton per day capacity mill had only 35 tons of ore available per day.

As was so common in the countless undertakings on Gold Mountain, for every two dollars raised from the stockholders, only one went into the development of the mine. The rest went into the pockets of the promoters.

With the collapse of the two giant enterprises on the mountain, the power plants on Fish Creek and Clear Creek became pawns in a power struggle between the Teluride Power Company and the new owners of the two mining companies. The Teluride's primary concern was to prevent these plants from falling into private hands to compete with them.

A lawsuit followed between the Teluride and the new owners. The latter failed in their bid to buy the plants, and they were bought by the Teluride Power Company.

In the powerplant settlement, the Teluride agreed to pay the Salt Lake Hardware and Fred Tilton a total price of $40,000. When offered a certified check for his share of the settlement, Tilton refused to close the deal until the check had cleared the bank. He cited the financial condition of the bank as his reason. In later years, the story was distorted to indicate that Tilton had refused to accept the check and had demanded gold.

The Teluride soon realized that they could not economically use these power plants, and they then sold them to the Annie Laurie Gold Mines. The lower plant on Fish Creek was destroyed by a flood and by vandals in 1909. This same flood also destroyed the road to the upper power plant. This fact protected it from vandals, and it continued to operate until 1911.

Fred Tilton came West from the state of Maine in 1880 as a carpenter. He worked on the Canadian Pacific Railroad and got a contract to build a railroad station, loading docks, and freight sheds. Later, he moved on to work for the Santa Fe Railroad in Arizona and New Mexico. Tiring of the railroads, he bought a large herd of sheep (approximately 2,200 head) and drove them to the Kaibab Forest in northern Arizona where he had obtained grazing rights.

In the lawless West, at the time, grazing rights were sometimes only as good as a man's ability to defend them. A Mr. Stafford, a longtime local resident, resented newcomers and twice threatened

Tilton and his partner, Arthur Weymouth. The dispute came to a head when Stafford again came to the Tilton camp and threatened to kill Tilton. He drew his gun, but Tilton was faster. Stafford fell dead. A hearing ruled the incident as justifiable homicide. In later years, Tilton took great pride in the fact that even though he was not a Mormon, he received many letters from Mormon well-wishers, including a Mormon bishop, stressing their belief in the justification of the shooting.

Tilton sold his sheep and moved to Salina, Utah, to try his hand at cattle raising. There he met and married Adelaid Schroeder and moved to Provo where, with two partners, he opened the Provo Commercial and Savings Bank and became its president.

In 1905, Tilton sold his 34 percent interest in the bank and returned to Salina. He purchased one-half interest in a grocery and general merchandising store in Salina; then seeing the growing opportunity in booming Kimberly, he purchased one-half interest in a store there with Max Krotki.

Contrary to many stories and legends, Tilton was not the coarse, rough man the stories depicted him to be. The killing of Stafford branded him and the reputation followed him.

After the collapse of Kimberly in 1908, hundreds came back to the mountain to "claim property" which was spread over the 2,000 acres. Items disappeared by the wagon load. In an effort to stem the flow and salvage what he could of his investment, Tilton blasted the roads closed from the Holland Mine in Sevier to Mill Creek, forcing all traffic to come through Kimberly proper. The law had left town and strong arm tactics, or threat of them, was the last resort in survival, Tilton felt. His reputation grew.

CHAPTER 6

The Quiet Years

The ghost-like quiet after the 1908 crash contrasted sharply with the bustling life of previous years on the mountain. Silent houses stood stiff and empty on naked hills stripped bare of vegetation. The smoke-laden air had vanished, and the large mill set sullen against a clear-blue sky—a monument to the failure of man.

The new owners of the Gold Mountain complex licked their wounds and collected their assets. The Salt Lake Hardware, with 80 percent of the stock, was in no position to manage this distant acquisition. Tilton, with 20 percent of the company, was on the scene and was capable of such management. An agreement reached between the Salt Lake Hardware and Tilton gave Tilton a $250-per-month salary from the Salt Lake Hardware to stay and manage the holdings until such time as the property might be sold. Tilton established his home and base of operations in the lodge.

Some of the visible ore in the Annie Laurie still assayed fairly high. Tilton reasoned that with the lower overhead of a small crew and the more efficient milling technique of the Sevier Mill, the combination could make the operation economically sound. A study was made of the ore prospects in the Annie Laurie by a Charles Durham. From this study, Tilton drew up this plan:

1. Keep all tunnels open except Number Five.

2. Block out enough ore from the upper levels to show that the mine had a future.

3. Accumulate 10,000 tons of ore of at least $10 per ton value.

Over the next three summers, the 10,000 tons of ore mined in the Annie Laurie wound its way over the treacherous roads to Sevier in large horse-drawn wagons. This was one of the few mining operations in the area to make money during this period. The profits ranged from $7,000 to $30,000 per summer.

The Surprise Gulch, located between Kimberly and Sevier, offered the shortest route between these two points. In earlier days, excitement had soared over the prospects of the Surprise Mine. Very little of value was ever found there, however. When the Annie Laurie began hauling ore from Kimberly to Sevier, the Bank of England claim was filed in the gulch, presumably for the sole purpose of stopping the Annie Laurie from using the gulch as a transportation route. They then attempted to sell the claim to the Annie Laurie for an exorbitant price. When the Annie Laurie was able to circumvent the Surprise Gulch, they sold the claim for $400.

During this period, a gold brick valued at $3,000 disappeared from the mill. The disappearance remained a mystery for 20 years. Then the Salt Lake Hardware received a letter from an assay office saying they had bought a gold brick that had come from Kimberly. Again, this was proven by the silver and other metal content of the brick. Although the seller was identified, no proof of theft could be established due to the lapse of time.

In 1911, the visible ore was depleted. The ore reserves were unknown. Obviously, it would re-

Winter at Kimberly, 1928. Mill, center right; Lodge, center left; Upper Kimberly, upper center. Photo by Ed Tilton.

quire a large amount of money for exploration and development. The plan, then, was to prepare the company for sale. A small crew remained to keep the tunnels open and for limited exploration.

In 1913, an offer of $400,000 was rejected. World War I further strangled the struggling company. Cyanide became scarce and wages climbed. In 1915, an interested mining company sent an engineer by the name of G.B. Wilson to Kimberly to evaluate its possibilities. Wilson was in the process of getting a divorce and spent the entire summer on the mountain, much of it secluded in the Joe Nelson cabin at the head of Fish Creek. Very talented and alert, he prepared a very complete evaluation of the Annie Laurie Gold Mines. (See Appendix.)

Wilson believed that a large amount of ore still existed between Number Four and Number Five tunnels. Since Number Five was caved in, the network of tunnels between the two was full of water and opening Number Five to relieve the water would be costly. His report read in part:

"It can be stated here that the reason for the failure of the original company can, without doubt, be ascribed to mismanagement. The ore bodies are large, and the hanging wall is none too solid. Care and experience are necessary to hold the large stopes open. Attempts were made to work the stopes by a sub-level method which I should judge to be hopelessly adapted to the conditions. The company had paid out its earnings as dividends and was without funds. The inefficient management covered their failure by announcing that the ore was exhausted, and this has been generally believed to be the case until recent development work has disproved it." Wilson concluded his report with the statement, "I think the present owners have an exalted idea of its value, however if an option can be obtained at a reasonable purchase price under easy terms of payment, I would advise examining the mine exhaustively with a view to its purchase." He recommended that the purchase price should not be over $500,000.

The recommended offer was presented to Mr. B. M. Bauers of the Salt Lake Hardware. Bauers countered with an offer to sell for $750,000. The offer was refused and haggling continued. Tempers flared and arguments raged between Tilton and Bauers. Letters flowed and prospective buyers came and went. Contrary to other stories, the $500,000 offer was the largest amount offered for the Annie Laurie Gold Mines.

Activity in the mines varied from short limited spurts to a complete standstill. Occasional leasers

Earl Mills wagon team at Upper Kimberly, 1930. Nothing was more precious to a freighter than his horses. "What would you do if you had a million dollars?" one was once asked. "I'd buy the best damn four-horse team on the road," he answered. Photo by Floyd Herring.

worked the various tunnels of the Annie Laurie holdings with negligible results. Charles Shaw was one such prospector whose unhappy life ended in tragedy. Well-educated, Shaw came to the mountain to be by himself as a result of a blighted love affair. He built a cabin alongside Grasshopper Creek just north of Lower Kimberly's business district. An outside income kept him comfortable, and he spent his summers prospecting and occasionally leasing one of the Annie Laurie Claims. Deep snows of winter drove him to the valley, but he returned each spring to the enchantment of the mountain. In 1913, he returned to the mountain early and leased the Bluebird Mine. On February 26, he left the Bluebird hiking down the snowy trail to Kimberly. He never arrived. One-quarter of a mile below the Bluebird, a huge snowslide leveled a path 100 feet wide through the quaking aspen trees and across the road. The *Richfield Reaper* recorded this item on May 13, 1913:

"The remains of Charles Shaw, a well-known miner of Kimberly who disappeared on February 26, have been found. Mr. Shaw came to the camp four years ago, and has worked there ever since, being employed by the Sevier Mining and Milling Company, and by others. He was highly respected by all who became acquainted with him.

"At the time of his disappearance, a search was conducted by a party of 10 or 12 men for a period of ten days. They were convinced that he had left the mine where he was working on a lease and had taken the trail to his cabin. There had been

a number of snowslides on the trail, and though a thorough search of these slides had been made, no trace of the missing man was found. The Sheriff of Piute County was called upon and assisted in the search, but without avail. Since that time, a continuous watch has been kept up for the missing man and, on Tuesday, his snowshoes were seen protruding from the snow on the trail. An immediate search was made for the body, and it was soon brought to view, being covered by about 14 inches of hard, frozen snow. The remains were left as discovered until the sheriff could view them.

"As had been surmised, the unfortunate man had been killed by a snowslide. He apparently had been going uphill and the snowslide had met him, knocking him over on his back and smothering him. This supposition is made good by the fact that the body was found with the head downhill.

"Mr. Shaw has a mother and a sister living in Glen Falls, New York."

The Curnoe Brothers, Jack and Nick, were two of a close-knit group of Englishmen called "Cousin Jacks." Between assessment jobs for the Annie Laurie, they leased claims on their own and never lost hope of making that one great strike. Though fortune escaped them, they hovered on its brink on special occasions only to have the vein run out after a token treasure. Even their attempt to raise a garden in the rich mountain soil ended in failure. Though the vegetables sprang from the ground under the warm stroke of the summer sun, the season cut them off short and nothing but therapy was realized from their effort. Their legendary sense of humor was undaunted, however.

Once, a huge rock tumbled into a main shaft deep in the Annie Laurie and lodged part way down. Jack Curnoe and Floyd Herring climbed several hundred feet of ladders to the jammed rock. It hung precariously above their heads, its size filling almost the entire shaft. The question was what to do? The rock had to come out, yet any jarring might send it down on top of them. Jack turned to Floyd.

"Well, partner," he said, "let's go send Nick up here. He doesn't have any kids."

Jack's deviltry kept Nick vigilant. Once when Nick was preparing the evening meal, Jack came in and suggested that if Nick would go out and cut some wood, he would prepare the meal. Nick agreed. Within minutes, Nick came in rubbing welts on his arm. "There's a damn hornet's nest out in the woodshed," he exclaimed.

Jack continued working, expressionless at the stove. "Yeah, I know. That's why I didn't want to cut the wood."

As stated above, these English "Cousin Jacks" were a close-knit group. When Jimmy (the tough little Englishman who was scaled by his wife and squashed by a mule) died, someone remarked, "I liked Jimmy. He was a good worker."

Nick bristled. "Good God," he said, "Pete, the mule, is a good worker too. I liked Jimmy because he was Jimmy."

Three-quarters of a mile to the southwest above the Annie Laurie, Billy Johnston clung to his holdings on the mountain. He had a keen mind and almost unlimited energy. Though less than five feet tall, his size in no way restricted his digging. Mole-like mounds ringed his park-like cabin area and countless tunnels penetrated the hillsides. A cold, clear spring almost at his doorstep completed his mountain retreat. Very little paying ore ever came from his diggings, but he convinced the Annie Laurie Gold Mine management that there might be, and he sold several of his claims to them for several thousand dollars. Then he bought a hotel in Richfield and retired.

Another typical man of the mountain during this near-dormant mining period was Joe Jarrett. A drummer in the Civil War, he came to the mountain while still a young man. He staked a claim tucked away in the heart of the mountains over the hill about three miles south of Kimberly. He

Typical Gold Mountain cabin located midway between Number Four Annie Laurie and the Bluebird, taken about 1924. Photo courtesy of Elmo Herring.

built a sturdy cabin where he lived the greater part of his life. He could neither read nor write and tried to conceal the fact. It was said that he only bought canned or packaged groceries that he could recognize by pictures or past association. He liked the Curnoe brothers and brought his letters to them in Kimberly to have them read. His excuse was that he had lost his glasses. He more or less abandoned this deception each month, however, when he received his pension check. He then sought out Nick to have him witness his "X" in the check signature block. He attempted to build Nick's ego with, "Nick, you sign this, and you'll have your name in Washington, D.C."

Jarrett eventually tired of this imposition also, and he hiked to Kimberly to see Nick. "Nick," he said, "show me how you would write Joe Jarrett." Nick wrote Jarrett's name on a piece of paper then handed it to him. Jarrett looked at it quizzically, then stuffed it in his pocket. "Yeah, that looks pretty good," he said. Then he returned over the hill to his cabin.

A few weeks later, Nick hiked the three miles over the hill to see Jarrett. Upon his return, he

The Billy Johnston cabin is one of the few that have survived complete destruction. Johnston sold his claims and bought a hotel in Richfield which later became a well known landmark. Photo by Dean F. Herring.

reported, "Every piece of paper, loose board and log has writing on it just like a chicken had been scratching. And by God, some of it is beginning to look like 'Joe Jarrett'."

Jarrett valued his friends, and foremost among them was Jack and Nick. He was proud, however, and resented any assumption that he was not completely self sufficient. Once when his long absence from Kimberly alarmed the Curnoe Brothers, they hiked over the mountain to check on him. Jarrett was happy to see them and began to cut off some venison steaks. Then Nick suggested that they had become worried that he was sick since he had not been down for supplies.

Jarrett was furious. "Don't you think I've got sense enough to get out of here if I get sick?" he roared. He began to put the steaks away.

Jack, being a diplomat, suggested that he might have had an accident. Jarrett thought a moment. "Well, I suppose that could happen," he agreed, and he again began sawing on the steaks.

On his frequent trips to Kimberly, Jarrett was often observed browsing through a newspaper, usually upside down. His arguments with Ed Holmes, another Civil War veteran, were legend. Each doubted the other's tales of the war. Holmes, a watchman at the Sevier Mines, was described as a well-read man of the world, and he delighted in pitting his knowledge against the affable Jarrett's love of argument. Jarrett's verbiage and pronun-

ciation were described as "unforgettable." Jarrett's arsenal included a .44 caliber rifle. He explained that he used .44-70 ammunition for small game and .44-90 ammunition "for bars and sech."

One of the few phonographs on the mountain belonged to the very proud Joe Jarrett.

Jarrett routinely made the hike over the 10,000-foot high mountain to Kimberly until his late seventies. Typical of the mountain breed, he didn't trust banks.

During the late 1920s, Floyd Herring frequently visited Jarrett at his neatly kept cabin. Jarrett showed him a large tobacco can described as "the type with a bail on it." It was full of gold coins. "When I die," Jarrett said, "this is going to be buried with me."

In 1931, Jarrett became ill and hiked to Kimberly. Friends took him to the home of a sister in Monroe. His condition worsened and, in his delirious state of mind, he mentioned the gold buried under a tree. Without divulging more, he died.

During the years since, dozens of fortune hunters with metal detectors and shovels have combed the wooded hillsides around the Jarrett cabin. The gold has never been reported found.

The remoteness of Jarrett's cabin has discouraged all but the most hardy visitors since his death and, for that reason, this is one of the most intact of the old cabins on the mountain. The floor,

Joe Jarrett, a Civil War veteran, lived a solitary life as a prospector in this virtually undisturbed cabin three miles by trail over the hill from Kimberly. He died in 1931 leaving a legend of hidden gold. Photo, 1978, by Dean F. Herring.

however, has been torn up, apparently by seekers of the legendary gold. Two ominous indications suggest that someone did find it. An empty tobacco can of the type described as having contained the gold sets on the table. A message scratched on the table announces, ''All the gold has left, so keep out.''

Between 1924 and 1931, exploratory work moved at a snail's pace. The work force varied from three to ten men. Just keeping the tunnels open proved to be a never-ending task. The power plants in Fish Creek had long since been abandoned. The small amount of power needed at Kimberly came from the Clear Creek plant.

Floyd Herring joined the small work force in Kimberly in 1924 during the construction of a new power line. Ed Tilton described the new employee as exceptionally agile. Fred Tilton saddled him with the nickname of ''Chipmunk'' after watching him climb the poles and trees without the benefit of ''spurs.'' Herring became Tilton's right-hand man and remained in Kimberly until its final closure.

The skeleton crew, by necessity, were ''jacks of all trades.'' They maintained the tunnels replacing rotted timbers and lagging, and cleaning out the fallen debris. Slowly, the tunnels and stopes

Floyd Herring, long time miner at Kimberly, tells of Joe Jarrett's gold. Photo, 1974, by Dean F. Herring.

In 1978 after a half a century, the bed remained intact in Jarrett's remote cabin.

The torn up floor of the Jarrett cabin reflects a half century of vandalism, possibly searching for the legendary gold.

were extended in the never-ending search for gold. When the lumber and sawed timber supply ran low, the crew fired up the old steam boiler and sawed up a new supply. When the log pile diminished, they harnessed the teams, then chopped and hauled the new logs. When the mule's shoes wore thin, they shod them. Life was never dull nor routine.

This was a peaceful time at Kimberly. Lush green returned to the de-nuded hillsides. The remoteness and, as yet, primitive modes of travel kept the traveler and curiosity seekers away. To the privileged few who lived there, this was, indeed, a dream world. The three to four families who lived there roamed through a treasure chest of abandoned buildings, articles, mountain trails and remote cabins. The scores of deserted houses, cabins and buildings had gone almost untouched

A tobacco can with a bail, the type described as having contained the hidden gold, sets empty on the Jarrett table. A scratched message on the table top warns, "All the gold has left, so keep out." Photo, 1978, by Dean F. Herring.

since the exodus of 1908. Old magazines, calendars and newspapers cluttered almost every building. With most windows still intact, the wallpaper remained in good condition. This covering was glued to a white cloth fabric called "factory," which was tacked to the walls. With the coming of the depression, this cloth became a ready-source of material for much-needed clothing and household use. Piles of dishes, utensils, tables, and chairs set silent and unclaimed. Stacks of books still set on desks at the small schoohouse, and school records lay strewn to the four walls. Dr. Steiner's office was a particularly fine bit of treasure. Medical books and medicine bottles (many still full) lined shelves. The operating table set intact covered with rat dung. Pounds of beer bottle openers lay scattered throughout the saloons. Copper tubs, boilers and tea kettles were tossed by the winds. Huge four-oven wood stoves still set intact in the boarding houses. A hand operated potato slicer set clamped to a rat-dung covered table in the old Cummings boarding house. Coffee grinders and meat grinders were commonplace. Bedsteads and springs still remained in most houses.

The scavenging tourist or antique treasure seeker were unheard of in Kimberly. To the local residents, the items were junk, thrown away by those who had departed. Tragically, the true value

Slowly, but surely, the Joe Jarrett tunnel succumbs to time's elements. Photo, 1978, by Dean F. Herring.

of these treasures were not recognized and, bit by bit, it was thoughtlessly destroyed or disappeared. The author, himself as a boy of 10, broke hundreds of bottles in the doctor's office with a sling shot. Coffee grinders were toys until broken. Old cast-iron stoves were shattered in senseless vandalism. In some houses, old crank-type telephones still hung on walls. The phone system was still operational. A long ring alerted the Alma Christensen ranch eight miles away in Clear Creek Canyon. A connection could then be made to the outside world. One by one, the telephones were torn apart to obtain the intriguing magnetos and bells.

The roads and trails led the hiking adventurer to countless abandoned tunnels and cabins half hidden in the groves of trees, and almost untouched since the collapse of 1908. Ore cars and wheelbarrows set on mine dumps and near-abandoned cabins. Fallen water flumes and sluice boxes lay scattered. Deer, porcupine and chipmunks scampered through brush and trees. Pine hens slithered and bluejays scolded.

Through the summers, a huge sheep herd grazed across the Tushar Range. Moving leisurely, they drifted about a mile per week tended by two sheep herders who also became a part of the mountain legend. Leonard Abrhams, Wayne Day, Arthur Paxton and Owen Despain were among them. Every two or three weeks, the sheep herders moved camp to a new location following the meandering herd. Throughout the long summer, they seldom saw another human being. Visitors to their camp sparked wide grins to their bronze, wrinkled faces. A gift of home-baked bread brought whoops of joy and a guaranteed invitation to stay for supper. To the tired hiker, fried lamb cooked on a campfire and fresh-baked sour dough biscuits with butter rivaled a king's banquet.

The automobile had arrived on Gold Mountain and travel was easier, but the steep wagon roads and high thin air limited their use. It was common for the family to get out of the car and push to help the groaning vehicle over the steep grades.

The big day of the week was the trip to the valley 35 miles away to replenish supplies. A shopping spree in Richfield topped off with popcorn and ice cream was grand adventure. The trip back up the mountain from the sweltering valley heat was an adventure itself. Three miles below Kimberly at the Red Narrows, steam gushed from the

wheezing radiator without fail. An always-ready tin can hung on a convenient stump for filling the gasping unpressurized cooling system. Then the old car growled in low gear on to Oak Flat three-quarters of a mile ahead where the boiling radiator episode was repeated. Another three-quarters of a mile, at Water Cress, another enactment of the cooling process. The creek in Lower Kimberly near the old Lawson boarding house was the last cooling stop before destination's end at upper Kimberly.

With the coming of winter snow, the families moved to the valley where the children could attend school. The men who worked at Kimberly lived alone through the cold lonely winter.

Rats found a haven in the many empty buildings in Kimberly, and they knew no boundaries. Floyd Herring told of being awakened at night by the sound of scurrying rats in the ceiling above him. He developed a game of guessing the rat's location and shooting at the sound through the ceiling with a .22 pistol.

Once a week, he made the trip to the valley for supplies and a night with the family. At 4 a.m., the alarm clock sounded the end of his short visit at home, and the cold trek back to the mountain in the old windowless car began. Three- to six-foot snows stopped the automobile at the Red Narrows about three miles below Kimberly. From there, the travel was by skis or snow shoes while carrying the supplies on his back.

The air of a mansion surrounded the lodge. This was the Tilton home. Its massive rooms, huge fireplace, and modern conveniences inspired envy in those who visited it. The large fresh clipped lawn complimented the natural beauty of its hillside perch. Hand-made chairs fashioned from native birch set proudly on the huge porch that half-way encircled the building. A large stack of firewood covered most of the west side porch. The casual atmosphere lulled one into a dream-like world.

The proud and proper Fred Tilton lent dignity to the elegance of the lodge. He walked with the air of a gentleman—a man who knew where he was going—yet respected those around him. He earned respect rather than demand it. He liked children and paid the young boys five cents for each rat that they caught. He believed that a day's work demanded a day's pay. At the beginning of the depression when B.M. Bauers, of the Salt Lake Hardware, implied that Tilton was paying his help too

A sheep herder's camp. The aroma of frying lamb chops drives the tired hiker onward upon nearing a Gold Mountain camp. Visitors were always welcome.

Sixteen pound hammer and plate used for smashing ore samples.

Floyd Herring and Ervin Bridges arrive at the sawmill with a load of logs, 1924.

Steam powered sawmill at Number Four tunnel in 1927. Walt Johnson, front.

Blacksmith & snow sheds - Number Four Tunnel 1929

Smoke stack reached skyward from the steam boiler in the combination boiler room and blacksmith shop at Number Four tunnel. Steam boiler provided steam for sawmill (not shown) on left. 1929 photo by Floyd Herring.

Boiler room and snow shed complex lay in ruins, 1978.

much and suggested that he could send him all the miners he could use for one dollar per day, Tilton responded, "Send them down and I'll send them right back. I've got good men here."

During the summer of 1927, an unusual number of range cattle found their way into Kimberly and milled around the lodge. The indignant Tilton complained first to the cattle owners, then to the sheriff; but the complaint fell on deaf ears. The highly-manicured lodge lawn began to look like a barnyard. One day, after a particularly loose-boweled cow had saturated the steps to the lodge, Tilton's oldest son, Ed, shot the cow, then dragged the carcass over the bank with a horse. Tilton refused to pay for the cow, claiming she was trespassing.

A short time later while shopping in Richfield, Ed was stopped by the sheriff and fined $50 for

The Alma Christensen Ranch in Clear Creek Canyon was the telephone terminal and the gateway to Kimberly. Standing (L to R): Neal Julander, Unkn., Alma Christensen. 1928 photo by Floyd Herring.

Winters were lonely for the men of Kimberly during the 1920s through 1932. Families moved to the valley where children could attend schools. Photo by Floyd Herring, 1928.

A Christmas tree from Kimberly is brought by sled to the Red Narrows, the stopping point of the auto in winter. From there, the tree trek was on skis and snow shoes.

operating a California-registered automobile while residing in Utah. Sensing that the arrest was retaliation for the cow incident, Tilton was infuriated. He wrote a letter to the Richfield Commercial Club outlining his grievances with the city. He said, in part, "We have been spending over $10,000 per year in this community. Much of this money has found its way to your townspeople in wages and to your banks. It seems to me, we are entitled to the consideration of other citizens." To add emphasis to his indignation, he then withdrew his entire bank account from the Richfield bank and transferred it to the bank in Salina, 20 miles farther north. To further drive his point, he refused ever again to spend one dime in the city of Richfield. Each week, he drove the extra 20 mile to Salina to do his shopping. He also arranged to have a large store in Salt Lake send staples to Kimberly on a regular basis.

The law in Richfield was not the only branch that plagued the Tiltons. When Ed Tilton found two men breaking into the assay office, he pulled his revolver, took their rifles away from them, then sent them on their way. A short time later, the Tiltons were issued a warrant to appear in court on January 28, 1928. Tilton wrote the judge, "Our acts resulting in this charge were stopping two armed men, Harold Nielsen and Wendell Peterson, in our dooryard on October 30th and disarming them after they had been forbidden to trespass by appropriate signs; and after this, they had broken into, and ransacked the Sevier Mill." Tilton lost the case, and Ed was put under bond for one year. But, in Ed's words, "This abruptly stopped our problem with visitors looking for 'goodies'."

The problems were not stopped entirely, however. The town of Circleville came to realize that they needed a jail in that community. A meeting of the people produced a possible solution. Since the bankruptcy at Kimberly, the strong jail had set idle and was deteriorating. A committee was appointed to bring the jail to Circleville. The crew, led by Walter Crow, left for Kimberly with a four-horse team and wagon. They arrived at the

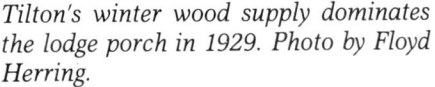

Tilton's winter wood supply dominates the lodge porch in 1929. Photo by Floyd Herring.

Snow covers the neatly manicured lawn of the lodge, the residence and office of Fred Tilton, manager and part owner of Kimberly. 1928 photo courtesy of Ed Tilton.

jail and began to unload their equipment. Tilton was on the scene in minutes demanding to know their business. Crow explained their mission and Tilton explained his. He suggested that they should leave the mountain bright and early the following morning without the jail. Tilton's suggestion was heeded.

In 1915, the lodge was the scene of an attempted rape. A young engineer and his nephew arrived in Kimberly, presumably to look for alunite, much needed for the war effort in progress. They stayed in one of Joe Nelson's cabins at the head of Fish Creek through the better part of the summer. The engineer was aware that the Tiltons had a housekeeper at the lodge, and upon hearing that the Tiltons were out of town, he hiked down to the lodge to visit the housekeeper. He found her alone and, in his attempts to force his attentions upon her, he became rough. The housekeeper picked up a piece of stove wood and struck him in the groin. When the Tiltons returned, Fred and Ed saddled their horses and headed for the Nelson cabin. The prospectors were gone when they arrived and were not seen on the mountain again.

Tilton had two sons, Edward G., the oldest, and Paul. Ed was seven years old when the Tiltons first moved to Kimberly. He spent his summers in Kimberly until 1931. "Following Dad was my obsession," Ed wrote. "Over the surface, through the tunnels, up and down ladders into raises and stopes—I was there ahead of Dad on the ladders so that I could be caught if I fell. I was taught to be a listener. Dad's quote, 'The fellow who talks all the time gives his brain no chance to pick up knowledge.' "

Ed was a studious young man. As he grew up, he spent his winter months in San Francisco going to school, then returning to the mountain in the summer. In his free time, he roamed the mountains on foot and horseback. When he could be of help, he took on odd jobs for his father, including surveying, running assays, repairing equipment, mule skinning, and horseshoeing. It was this latter chore that Ed told about with relish.

The miners and workmen liked Ed but delighted in joshing the Boss's son. When Ed decided to replace the shoes on Pete, the mule, the workmen smiled to themselves and needled him

Fred Tilton, General Manager and part owner of the Annie Laurie Gold Mines, inspects progress deep inside the mine. 1930 photo by Ed Tilton.

with skeptical remarks. Due to an illness which was not diagnosed for many years, Ed was not particularly robust nor athletic. Pete's intelligence was exceeded only by his stubbornness. Being his trusted friend was a prerequisite to being able to install new shoes on him. If he wasn't able to kick his "tormenter," he planted his four feet solidly on the ground and only herculean force could raise his foot. Pete didn't like Ed.

Refusing to admit defeat under the snickering verbal barbs from the workmen, Ed fashioned a harness under the belly of the obstinate mule. Then he led the mule into the blacksmith shop and, with a block and tackle, raised him bodily off the floor. The helpless mule relaxed and the shoes went on.

That the job was well done could not be denied, but the razzing intensified. "Only a boss's son could get away with something like that," they scoffed. "Why didn't you shoe him like a man?"

The story should have ended there, but there was a sequel. Ed returned to San Francisco in the fall, and the parting remark from Bill, the black-smith, was: "Well, by next summer you'll grow up enough that you won't have to use the belly band on the mule."

The next summer, Ed returned, and when he strode into the blacksmith shop, there on the wall was his mule's belly band. But most mystifying, it had been reinforced. Ed's inquiries about this strange turn of events was met with blank shrugs. No one seemed to know, or would not admit knowing anything about the reinforcement.

Eventually, the truth surfaced. Bill, the "he-man" blacksmith, had tried to shoe the mule, and was kicked across the shed. He had been hospitalized for three weeks. There was no more said about the belly band.

Mules were vital to the Annie Laurie. Their size (smaller than a horse, yet nearly as strong) made them adaptable to the low narrow tunnels. They are not as excitable as horses. They are stronger and more manageable than burros. Their intelligence is legendary.

Pete, a long-time favorite mule at Kimberly, worked like a machine. He picked his way through

Steam powered sawmill at Number Four tunnel. (L to R): Steve Baker (uncle of Ivie Baker Priest), Ed Tilton, Alma Christensen, Floyd Herring, Neal Julander. 1928.

the maze of tunnels flawlessly. Pulling a string of four to five ore cars, without prompting, he stopped with the lead car directly under the ore chute. The "mule skinner" (handler of the mule) opened the ore chute to fill the car. When the front end of the car was full, the mule skinner spoke gently. "Little bit, Pete." Pete then moved the car exactly the right distance for filling the aft section of the car. When that car was full, a soft verbal command moved the second car into place.

On the long haul from the tunnel, several down slopes caused the ore train to speed up, overtaking the mule. At the appropriate spot, Pete backed his rump against the lead car and braced his belly against the timbers on the side of the tunnel to slow the cars to a controllable speed.

Pete died of old age after 20 years in the mine.

Tragic accidents struck mules in the Annie Laurie as well as men. Andy replaced Pete at the head of the ore train. Soon afterwards, he lost an eye in an accident in the pitch black tunnel, but he continued at his job until the other eye began to fail.

More tragic than Andy's fate was that of Carrie. Carrie's intelligence rivaled that of Pete, and her gentle temperament bred affection. When not straining at the ore cars in the mine, she served as a saddle mule on the mountain trails. She was a familiar sight to all on the mountain and all loved her. Late one night, Ronald Ross, the mule skin-

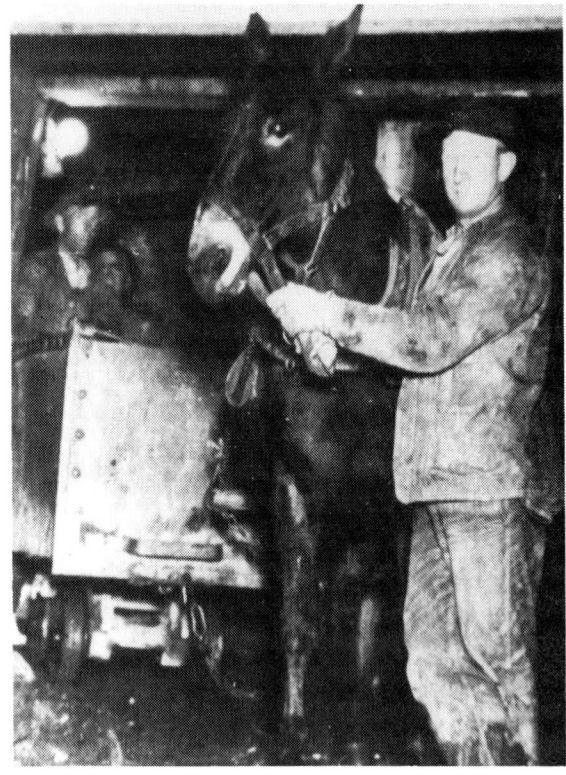

Carrie, the mule, deep inside the mine, 1930. Mules were indispensable to the mining industry.

Al Mills, teamster for the Annie Laurie, arrives at Number Four tunnel snow sheds with a load of hay from the valley. 1930 photo by Floyd Herring.

Merlin and Earl Anderson, Upper Kimberly, 1929.

ner, called Floyd Herring, the mine foreman. One of Carrie's tug straps had come disconnected and hooked under a rail stopping her short. The momentum of the heavy ore cars overtook the helpless mule and crushed her. The task of removing the carcass of an 800-pound mule in the tight tunnel seemed overwhelming.

"We'll probably have to quarter her up and haul her out in an ore car," Herring suggested.

"Oh God, I couldn't do that to Carrie," the distraught mule skinner groaned.

Carrie's loss cast a gloom on the mountain.

The depression of 1929 and 1930 added new pressures on management to sell the Annie Laurie Gold Mines. Partially to make the property more attractive to prospective buyers, a new power line was built from Sevier to Kimberly. This allowed Teluride power to flow from the power lines in Clear Creek Canyon. Also in 1930, a mining engineer by the name of Earl Young, who represented a prospective buyer, made a detailed report on the company's production history and prospects for the future. It was his opinion that since very little quality ore was ever mined in Sevier, any existing bodies would not persist in depth. He estimated that 450,000 tons of ore had been removed from the Annie Laurie. He believed there was another 350,000 tons to be taken, largely between Number Four and Number Five Tunnel. Because of the 350-foot vertical distance between

Corrals — upper Kimberly   1930

The stables in Upper Kimberly, 1930.

these tunnels, it was his opinion that an intermediate tunnel, Number Four and a Half, already under construction, was necessary for economical mining. The cost of reaching the hoped-for ore bodies would be high. Reopening Number Five Tunnel was not recommended unless Number Four and a Half proved successful.

Young's report reflected that the Annie Laurie Gold Mines Inc. consisted of 145 patented and 25 unpatented claims. The corporation was capitalized for 3,000,000 shares at $1 par value with 1,810,000 shares issued. The asking price for the company was $1,357,500. The prospective sale was not made known to Tilton. Bauers of the Salt Lake Hardware did agree to buy Tilton's interest, however. This agreement, which Tilton accepted, provided for a $50,000 cash settlement and $14,000 to be paid over a four-year period. It was also agreed that Tilton would stay on as manager at a salary of $350 per month.

Bauers again failed to consummate the sale of the company.

The country staggered under the weight of the depression and the skeleton crew all but ceased to exist. It semed that the mighty mountain had shaken free of man's encroachment, and nature would once again reclaim its soil.

Mine foreman's house in Upper Kimberly was one of only four boasting of a bathroom in Kimberly. Mrs. Floyd Herring on steps. 1928.

CHAPTER 7

Kimberly Booms Again

In 1933, the relative quiet that Kimberly had known was broken. A Mr. Alfred Bell negotiated a lease for the Annie Laurie. Investors, shaken by the deep depression, sought refuge in the gold of the mountain and, again, threw money into exploration. Miners were hired and reasonable bodies of ore between Number Four Tunnel and the Bluebird were exploited. The ore below Number Four was ignored due to the high cost of recovery. A new mill was built near the mouth of Number Four Tunnel adjoining the snow shed complex.

The new management left little for Tilton to do and friction developed. A new modern house was built and moved to Kimberly near the lodge for management's use. The lodge, which had been the Tilton residence for years, became a bone of contention. Management offered, or forcefully suggested, that Tilton should move into the new house leaving the lodge for the exclusive use of the new management. But Tilton was a proud man. The lodge and Kimberly were his life, and he would have no part of the plan. His advancing age weakened his will to fight and he left Kimberly forever. He died in Oakland, California, in 1941.

The new mill was small compared to the huge old mill. A rock crusher crushed the ore to a uniform size, then a conveyer belt carried the ore to a ball mill. The ball mill, revolving like a huge concrete mixer with several 20-pound cast iron balls inside, pulverized the ore into a fine sand. A new floatation process eliminated the costly and long-outmoded heating and drying procedure.

The mine, too, was modernized. Electric mules replaced the animals in the mines. The drone of the huge air compressor rolled through the hills day and night, pumping air to the busy jackhammers throughout the tunnel complex. Electricity replaced the steam engine at the sawmill.

Logging crews again ripped at the forests but on a smaller scale. Mining crews worked the tunnel complex from Number Four Tunnel to the Bluebird, and the honeycomb grew. Within a year, all of the livable houses bustled with new life. Automobiles ground their way over the narrow roadways, and roads were upgraded to accommodate them. Housing, water and lights were rent free. Upkeep of the houses was occupant responsibility. Building materials were free for the tearing down of the more deteriorated buildings.

One store, the old Kimberly Merc, opened in Lower Kimberly stocking groceries and commonly used commodities. The advent of the automobile eliminated the need for more.

The extremely low wages brought on by the depression was the element that allowed the company to exist. The average miner and mill man received $2 per day. The foreman's salary was $90 per month. Even at these wages, men clamored for jobs.

Almost no new buildings came to Kimberly during the new boom of the thirties. The new house near the lodge, the new mill (built almost entirely from scavenged material on site), and sanitary

Stopped by the snow, cars parked in Lower Kimberly. Kimberly Merc,
owned and operated by Fred Tilton, center left. Lawson Hotel on right.
1935 photo by Floyd Herring.

Mr. and Mrs. Lou Hardy carrying water from the creek. Old Lawson Hotel in background.
1935 photo by Floyd Herring.

outhouses were the exceptions. The latter, commonly referred to as "streamlined outhouses" or "Roosevelt monuments," were the result of a project in the Work Program of America (WPA) during the Roosevelt Administration. Each house in Kimberly was appropriately updated at no expense to the occupant.

Garbage disposal remained unchanged from the early days of Kimberly. Each household dumped its garbage on the downhill side of the house among the trees.

Refrigeration never arrived on the Kimberly scene except at the one and only store to open, the lodge and the new manager's house. A dugout

cellar furnished cool storage for perishables for many houses. Butchered deer or sheep hung in many woodsheds. During the warm days, a sheet or tarp wrapped around it kept the meat cool and deterred flies. When it became apparent that the meat would keep no longer, the remainder was canned. In Lower Kimberly, the old abandoned "ice tunnel" came into use again as a cooler.

Throughout the summer, a stream of peddlers' trucks crept up the steep mountain roads seeking market for their goods. The author, as a boy, remembers meeting these trucks in Lower Kimberly at one of their many stops. He, along with friends, hitched rides on the back of the slow-

Above: Miners eat lunch on pile of stacked timbers at the Bluebird Mine, 1935. Front (L to R): Rolo Ross, Grant Ross, Jimmy Stokes. 2nd row: unkn., Mac Sly, Glen Bridges. Rear row: unkn., unkn., unkn., Del Bridges, Cal Sly.

The Bluebird tunnel (left) at the 8,900-foot level linked with the Annie Laurie tunnels.

moving trucks to Upper Kimberly. Then, as the truck groaned up the steep grade, assorted fruits and produce departed from the truck into the bushes alongside the road to be picked up later.

The author also remembers the "powder magazine," an old abandoned blacksmith shop used by the mining company to store dynamite and black powder. Tons of explosives, dynamite and black powder reached to the ceiling in the unlocked building. Cans of black powder set open on the floor. Countless pounds of explosives found their way to the remote reaches of the mountain where super fireworks rocked them. Town folks speculated but never openly guessed the source.

This building had an attic of sorts and, being secluded, served as a "hideout" where an occasional cigarette could be enjoyed. It was quiet there and parents would never suspect. In the smoke choked quarters, no thought was given to the tons of dynamite and black powder spilled like sand on a beach visible below through the two-inch cracks.

Being raised in a mining environment, it was

Keeping the roads open. Sim Peterson and Randall Herring, 1935. Photo by Floyd Herring.

Clearing the road between Upper and Lower Kimberly, 1935.

natural to dig tunnels like the big miners do. The author's particular tunnel was born in a heavy stand of quaking aspen trees about 100 yards above the new mill. When digging became hard, dynamite was the natural solution. A three-foot hole was drilled and dynamite tamped into the hole. The fuse was lit and the premises evacuated. The entire mountain trembled and three amazed boys hurried back to the blast site. They were more amazed to see 30 men from the mill rushing toward them. This turn of events turned awe into panic and like stampeding chimpanzees, the site was abandoned. The men from the mill, seeing the fleeing juvenile crew, feared another explosion and took cover back in the mill.

Carbide was another great adventure. The candles used by early-day miners had long since been replaced by the carbide light. These lights burned acetylene gas formed by water uniting with carbide. The upper part of the lamp was a water reservoir. The lower section contained a supply of carbide. The water, controlled by an adjustable valve, dripped slowly into the carbide. The resulting gas ejected through a burner jet. A flint striker produced a spark to ignite the gas.

The amazing qualities of carbide invites experimentation. As stated, when water is added, it froths and boils getting very hot. A favorite subject for experimentation was a half-asleep workman sunning himself against a building. A few lumps of carbide slipped into his pocket, then dampened with water, usually brought him to full capacity. A well-planned escape route was a must.

Exploding beer bottles with carbide was another exciting pastime. Assured survival required two people. A few lumps of carbide was poured into the bottle and water added. One person held a cork and the other a board. The cork was positioned at the bottle opening and the board drove it home. A hasty departure was of paramount importance.

The social centers so prominent in early Kimberly failed to return with the new settlement. The advent of the automobile caused less dependency upon neighbors, and social cliques remained

Sawmill at Number Four tunnel. Reed Christensen on left, Leo Staples at saw. Roy Thornton, center right. 1934 photo by Floyd Herring.

Taking a break at Number Four tunnel. Sonny Nielson on left, others unknown. 1934 photo by Floyd Herring.

In 1973, only the fireplace remained where the lodge had once stood. Photo by Dean F. Herring.

small. The baseball team was an exception. The Sunday game between the millworkers and the miners at the old reclaimed ball park became the big event of the week.

Once again, winters found Kimberly a beehive of activity. The two-room schoolhouse found new pupils, though few in number. Ten to 15 pupils represented grades one through eight. Most families returned to the valley for the better schools in winter.

The network of trails etched through the wooded mountainsides to far-flung work sites were buried in snow. Dozens of plodding boots trampled trenches over them like black threads across winter's blanket. Again, blue smoke from wood stoves curled from chimneys, and the drone of the giant compressor at Number Four Tunnel rapped the still mountain air. Now and then, the crackling exhaust of a caterpillar echoed across the canyon as it cleared the snow clogged roads.

Inside the houses, scenes varied very little from the early days of Kimberly. A roaring fire crackled in the wood cookstove and heater. Wet boots lined the floor around the stove and heavy woolen socks hung to dry on makeshift clotheslines behind the stove. A table and chairs set on one side of the bare kitchen. A few shelves on the wall served as a cabinet for dishes. A steaming teakettle on the stove supplied the hot water for most needs of the family. The cold water ran constantly in the kitchen sink to keep it from freezing.

Radios began to appear in the otherwise austere living rooms. A couch and a few chairs completed the living room furnishings. Electric washing machines made their debut, but telephones were the exception rather than the rule.

A few "free lancers" still prospected and leased holdings on the mountain. Notable among them were Bill Sawyer and Perry Miller who leased the Hastings Tunnel between the Bluebird and Number Four Tunnel. Pursuing veins of ore judged too small for an operation the size of the Annie Laurie, they clawed a small fortune from this tunnel allowing both to leave the mountain much better off financially than most.

Those men who braved the winters by themselves came home to frigid houses and spent the evening chopping wood and stoking fires. A self-prepared meal climaxed his day, and he buried himself in a mountain of blankets.

In the mines, the rambling tunnels and stopes grew in length and complexity. Drills, picks and shovels chased each new vein showing promise. With each disappointment, the doggedly determined crews backed off and sought new direction. New mining engineers scoffed at the early-day miner's ability to follow the elusive veins. Floyd Herring, a mine foreman, had more respect for the oldtimers. At the end of a shift, one of his crew reported excitedly. "We ran into some stuff that looks pretty good today."

Floyd smiled. "Well, you'll probably run into an old tunnel tomorrow." The following day proved him right. The oldtimers had been there.

Wages improved slightly. Miners and mill men were paid $90 per month; the foremen $120. But the price of gold did not change and the quality of the ore deteriorated. Rumors again ran wild.

The mine and mill were going to close. Frantic efforts failed to raise more money for continued exploration. In 1938, Kimberly collapsed for the last time. It wasn't a spectacular collapse. Employees were paid in full. There were virtually no businesses to wilt and die.

Like an old man who had struggled for a second chance at life, its life blood ceased to flow. It died gracefully with honor.

Available records reflect the following names employed during this period:

L.L. Anderson, Mucker
N.D. Anderson, Mill operator
Ben Andrews, Mill operator
Dewitt Bailey, Float operator
Steve Baker, Miner
Verle Baker, Store
Walt Baker, Miner
Alma Baird, Labor
Ross Baird, Miner
Clarence Barney, Timber cutter
Eldred Barney, Crusher
Frank Barney, Mucker
Larcel Barney, Labor
Lenard Barney, Miner
Loid Barney, Timber cutter
George Bertrams, Miner
Earl Borg, Truck helper
Carlos Bridges, Mule skinner
Del Bridges, Shift foreman
George Bridges, Mill operator
Glen Bridges, Mucker
Rex Bridges, Labor
R.L. Bridges, Float operator
Arnold Christensen, Mucker
Lee Christensen, Mill operator
Reed Christensen, Mechanic
Denzel Cope, Mule skinner
Jack Cramer, Miner
Allen Crosby, Mule skinner
Champ Cuff, Laborer
Ferris Draper, Miner
R. Draper, Miner
Karl Dunn, Mucker
John Elder, Mucker
Ray Elder, Miner
Lester Epling, Miner
Merlin Epling, Miner
Frank Ernst, Mucker
Aldo Francovich, Laborer
Heaps Guardner, Timberman
L.B. Hadden, Miner
Arrol Halladay, Repairman
F. Hansen, Mill operator
Louis Hardy, Miner
J.B. Haywood, Mucker
Van R. Haywood, Mucker

C.M. Hermansen, Mucker
Ned Hermansen, Mucker
Elmo Herring, Blacksmith
Floyd Herring, Mine foreman
Randall Herring, Miner
Don Hunt, Float operator
Lloyd Hunt, Mucker
Raymond Hunt, Miner
William Jaap, Tailings
Elden Jensen, Mill operator
LaGrande Johnson, Misc.
Urban Johnson, Truck driver
Lawrence Keller, Mucker
F. Kelsey, Misc.
Russel King, Miner
Ray Lavender, Sampler
Jim Lawrence, Misc.
Dean Leavitt, Mucker
Ether Leavitt, Misc.
Elmer Levi, Miner
Lester Levi, Carpenter
J.S. Mackey, Misc.
Hyrum Marquardsen, Carpenter
Claron McIntosh, Miner
Clyde Menlove, Assayer
Robert Morrey, Teamster
Charles Nay, Miner
Morris Nay, Mucker
F.D. Nielson, Carpenter
J. Grant Nielson, Miner
K. Nielson, Crusher
Paul Nielson, Miner
Peter Nielson, Unkn.
Ralph Nielson, Mucker
Howard Olcott, Mill operator
Robert Olcott, Misc.
Rulan Olcott, Miner
Billy Olsen, Miner
Melvin Olsen, Blacksmith
H. Clair Outzan, Truck Helper
R. Owens, Crusher
Fay Parks, Mill operator

Emerson Parry, Miner
Dale Parsons, Mucker
Charles Peah, Laborer
A.J. Pearson, Miner
Jay Peterson, Truck driver
Sim Peterson, Miner
Varlan Pinney, Sawyer
Frank Robinson, Mill operator
Culbert Ross, Misc.
Grant Ross, Miner
Rollo Ross, Miner
Sam Ross, Carpenter helper
Ruben Rudd, Miner
Doyle Shaw, Swamping
Art Shelton, Misc.
Bert Shelton, Miner
Ray Shelton, Mucker
Sidney Shipp, Miner
Cal Sprague, Miner
Cal Sly, Shift foreman
Mack Sly, Miner
Dean Sorensen, Electrician
Dick Sorensen, Miner
James Stacks, Miner
Leo Staples, Cat. Oper.
D.S. Stalks, Misc.
Bob Summerfruit, Timber cutter
Orin Tanner, Miner
Willard Taylor, Miner
William Thompson, Mucker
Elmo Thurston, Laborer
Lee Troutt, Miner
Clynn Utley, Misc.
Elgin Utley, Mill operator
Lynn Utley, Timberman
Mack Utley, Framer
Clayton Van Gordan, Swamping
Ralph Warenski, Mucker
Faust Whiting, Mechanic
George Wiley, Mucker
John Worsdell, Laborer

Annie Laurie mill, Lower Kimberly in background during its heydey in 1901.

Same view as above in 1973. Only the section where the old tramway entered the mill still stands.

The Annie Laurie mill viewed to the west in 1901. Photo courtesy of Elmo Herring.

Same view as above in 1973. Nature has all but reclaimed the raped hillsides. Only a small section of the mill still stands.

The last remaining house in Upper Kimberly now serves as a temporary store house for sheep herder supplies. Upon a return visit to Kimberly after an absence of 50 years, Josephine Pace observed, "Isn't it strange how a house gives up after those who loved her move away?" Photo by Dean F. Herring.

The end of the line. Just as the old Studebaker had reached its limits bucking the snow on the road to Kimberly, Kimberly too had run out of steam. Photo by Floyd Herring.

Once the main tunnel of the Annie Laurie system, the entrance is now marked by twisted timbers and fallen logs. Photo by Dean F. Herring.

The once bustling snow shed complex at Number Four tunnel is now a victim of plunderers and the elements littering the hillside.

One of the two remaining cabins in Sevier in 1977.

The rotting wood pipe protruding from the trail marks the route of the pipeline that brought water to the Sevier Mines.

The once familiar rumble of the heavily loaded ore cars has long since been silenced and their twisted hulks lay rusting in the creek below the dump of Number Four tunnel. 1973.

In 1978, only piles of debris show through the new growth where the Sevier mill stood.

The Sevier mill site. In 1978, only the rock foundations mark the spot where it once stood. Photo by Dean F. Herring.

Fifty years after Josephine Pace was forced to leave Kimberly after the collapse of 1908, she returned to visit the silent ruins. She wrote:

"I didn't realize how the years had embellished the things I remembered about Kimberly until I went back to find them. The houses had shrunk, been moved, or had fallen down in discouraging heaps. Isn't it strange how a house gives up when the people who loved it go away? The saloon was gone entirely when I went back, but in the pile of warped grey boards where I remembered the building had stood, I found an old flask turned purple by years in the sun. I liked to believe the bottle was opened on the last night the saloon operated, and someone who still had a few coins in his pocket drank a farewell toast and promised to come back when the mill started to turn again."

In 1942, less than 80 years after man's first onslaught on Gold Mountain, he stuck his last blow at Kimberly. The buildings and equipment were sold. Wrecking crews and trucks moved in. Wrecking bars and hammers ripped the proud old buildings from the lush mountainsides that had accepted them. Boxes of priceless old records trucked to Salt Lake City vanished in a jungle of bureaucracy and complacency. The priceless old four-oven cook stoves from the boarding houses brought a few dollars for scrap iron. Each item of value went to the highest bidder or was plundered by the quickest hand and disappeared in a cloud of dust down the mountain road to be lost to history forever. In a few short years, only scattered debris remained of this colorful bit of history. The priceless antiquity of one of the most fascinating towns of the old West was reduced to a few dollars worth of lumber.

The old mill fell under the wrecking crews' onslaught in 1943 and 1944. The Kimberly Merc, one of the last major buildings to go, was torn down in 1950. The proud lodge became a pile of meaningless boards in 1951. The old jail cell survived and, at one time, attracted curious visitors at Pioneer Village in Salt Lake City.

With the end of the resident watchman and the coming of better automobiles and trucks, came hordes of curiosity seekers to glean the last traces of value from building sites and garbage dumps. Some came with maps locating outhouses. In the days of prohibition, outhouses provided discreet disposal sites for whisky bottles, now valuable. What once had been a bustling community was now a scarred mountain and a memory to those who loved her. Seventy years after Woodruff Sylvester had delivered mail by horseback on Gold Mountain in 1903 as a 13-year-old boy, he wrote: "Kimberly! That name still gives me a thrill because of a mental picture I have of that once beautiful little mining camp. There isn't a spot on earth for which I have such fond memories. First, because I love nature, especially the mountains, and because I had the privilege of spending a lot of the golden hours of youth there." In 1973, he remembered, "The beautiful flowers—the columbines were there. It was a shame they tore down that nice old lodge and all those old buildings."

The clock ticks on and memories die. Time erodes man's traces. By 1978, only a sagging remnant of the old mill remained—a grotesque reminder of the recent past. Like carnage on a battlefield, scattered logs and warped grey boards give fleeting clues to a people's past. Logs from old stables lay twisted in heaps. The wail of the sawmill, the drone of the air compressor, and the rumble of the mill are long since silent. Below Number Four Tunnel's dump, battered ore cards lay rusting, half-buried in the bed of Mill Creek, their heavy rumble rolling from the mine forever still. Only the murmur of the creek breaks the quiet. The rattling harness of the wagon teams and the sound of the contented horses chewing oats at the devastated snow sheds and stables are but fading memories. Where the majestic lodge once stood, only the stone fireplace remains to mark the spot.

The entrances of the once bustling tunnels have long since caved in and waters of many seasons have sealed the gaping wounds with mud and silt. The mine dumps, once huge and bold, lay gashed and crumbling—fading with each season's storm. Hardy weeds sink roots into near sterile muck. New trees press hard seeking to conquer this bit of earth, blending once again into the mountain. New growth hides the scars where houses once stood. Trails have vanished and roads have lost purpose. The cries and laughter that echoed across canyons have been lost on the winds and empty silence reigns.

A lone house still stands in Upper Kimberly like a sentinel on a battlefield, giving shelter to an occasional sheepherder or adventurer.

The mountain's contribution to man will never be totaled. The gold and silver taken from its veins, if tallied, would be impressive. The investor's dollars would be staggering. In retrospect, hundreds of people who might otherwise have lived in poverty, found employment and a gratifying life there. Those who lived there gained priceless experience and unforgettable memories.

The low moan of winter's wind draws a cloud blanket over the lofty peaks. Huge snowflakes fall silently masking man's unsightly ruins. A tunnel caves deep inside the mountain's bosom; its rumble, like a growling entrail, is felt by only the mountain.

Wheels that once offered such promise to the men of the mountain rust in the creeks to become a part of it.

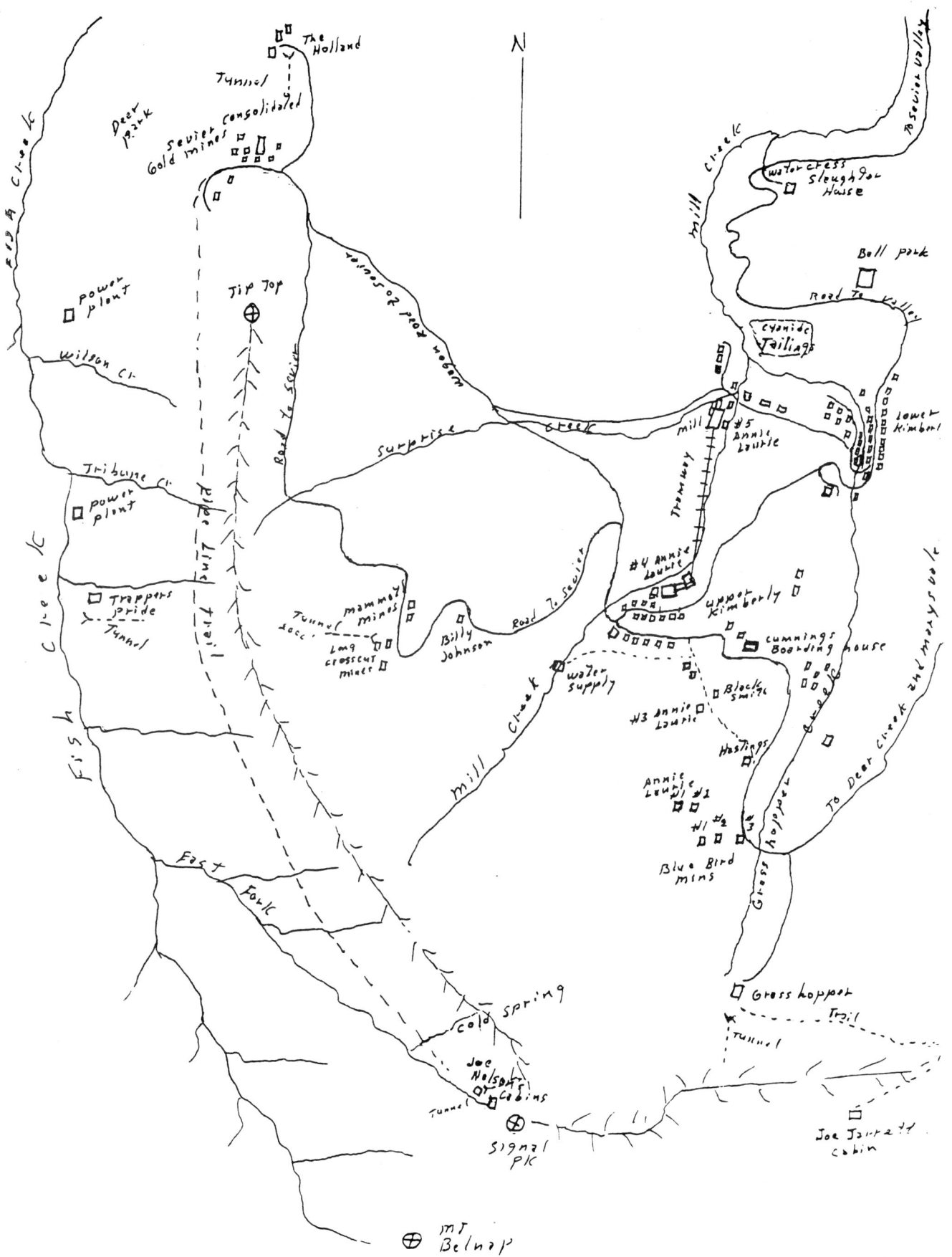

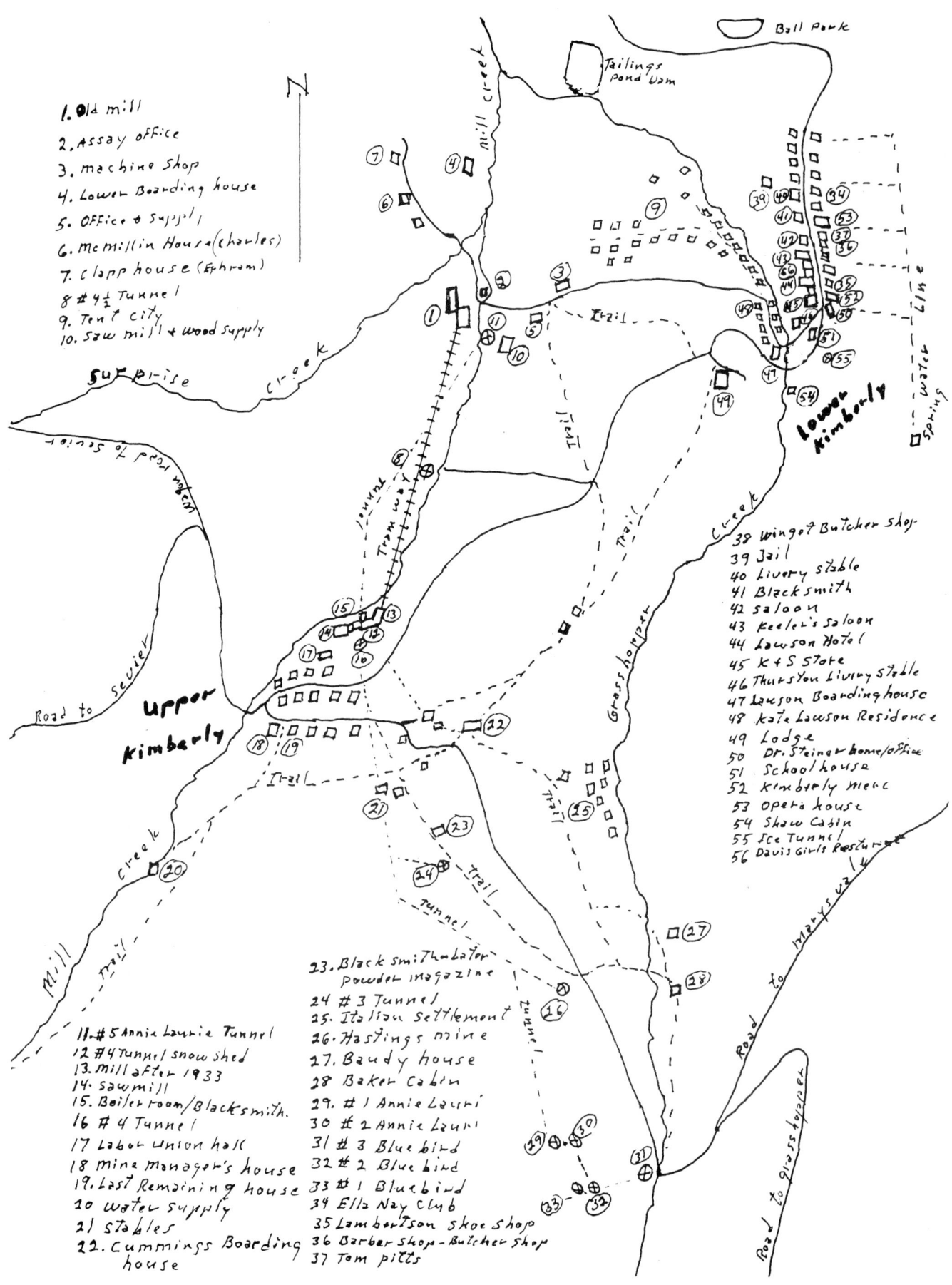

1. Old mill
2. Assay office
3. Machine Shop
4. Lower Boarding house
5. Office & Supply
6. McMillin House (charles)
7. clapp house (Ephram)
8. #4½ Tunnel
9. Tent city
10. Saw mill + wood supply

11. #5 Annie Laurie Tunnel
12. #4 Tunnel snow shed
13. Mill after 1933
14. Sawmill
15. Boiler room/Blacksmith.
16. #4 Tunnel
17. Labor Union hall
18. Mine Manager's house
19. Last Remaining house
20. Water Supply
21. Stables
22. Cummings Boarding house

23. Black smith later Powder magazine
24. #3 Tunnel
25. Italian Settlement
26. Hastings mine
27. Baudy house
28. Baker Cabin
29. #1 Annie Lauri
30. #2 Annie Lauri
31. #3 Bluebird
32. #2 Bluebird
33. #1 Bluebird
34. Ella Nay Club
35. Lambertson shoe shop
36. Barber shop – Butcher shop
37. Tom pitts

38 Wingot Butcher Shop
39 Jail
40 Livery stable
41 Blacksmith
42 saloon
43 Keeler's saloon
44 Lawson Hotel
45 K & S Store
46 Thurston Livery Stable
47 Lawson Boardinghouse
48 Kate Lawson Residence
49 Lodge
50 Dr. Steiner home/office
51 School house
52 Kimberly Merc
53 Opera house
54 Shaw Cabin
55 Ice Tunnel
56 Davis Girls Restaurant

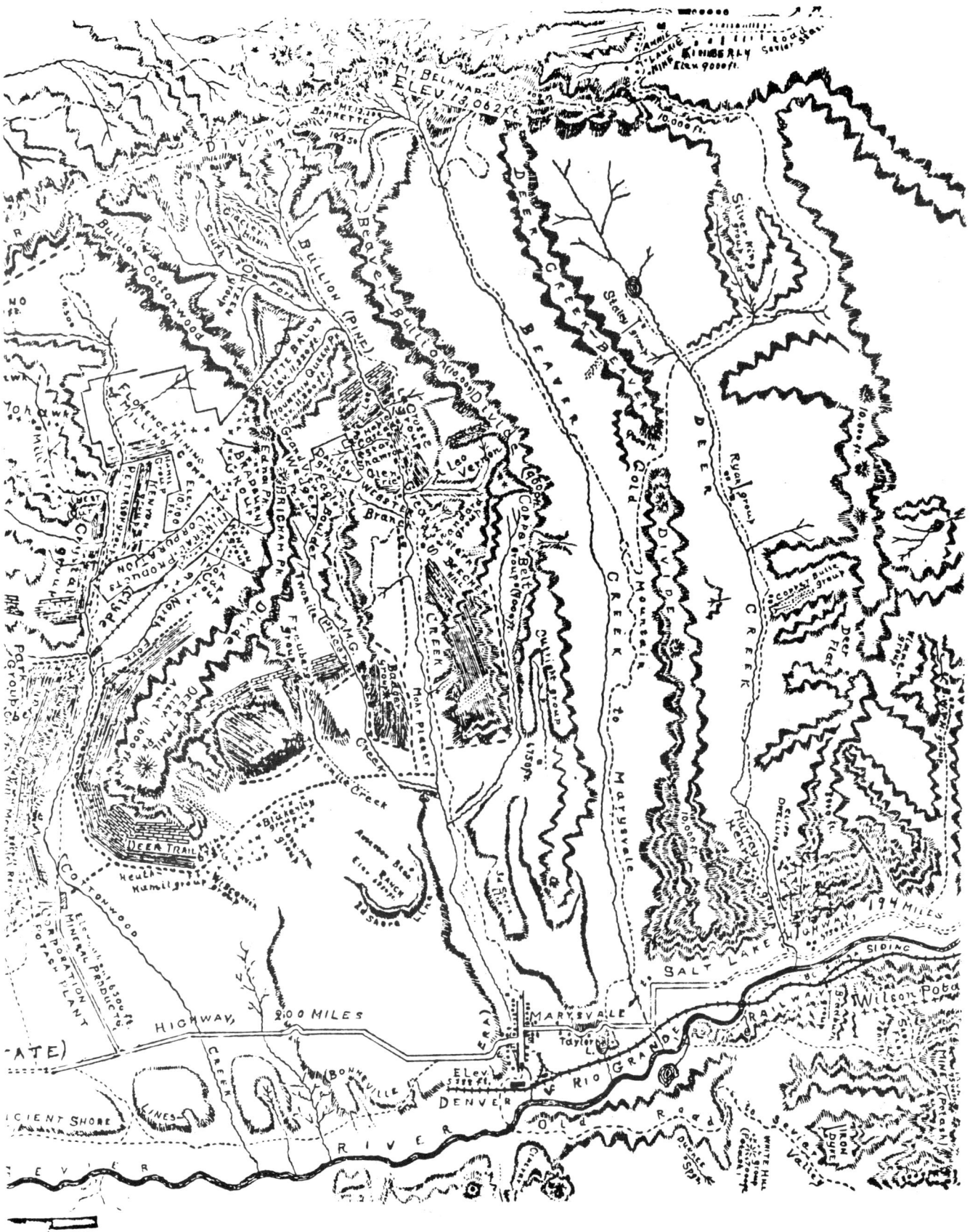

N

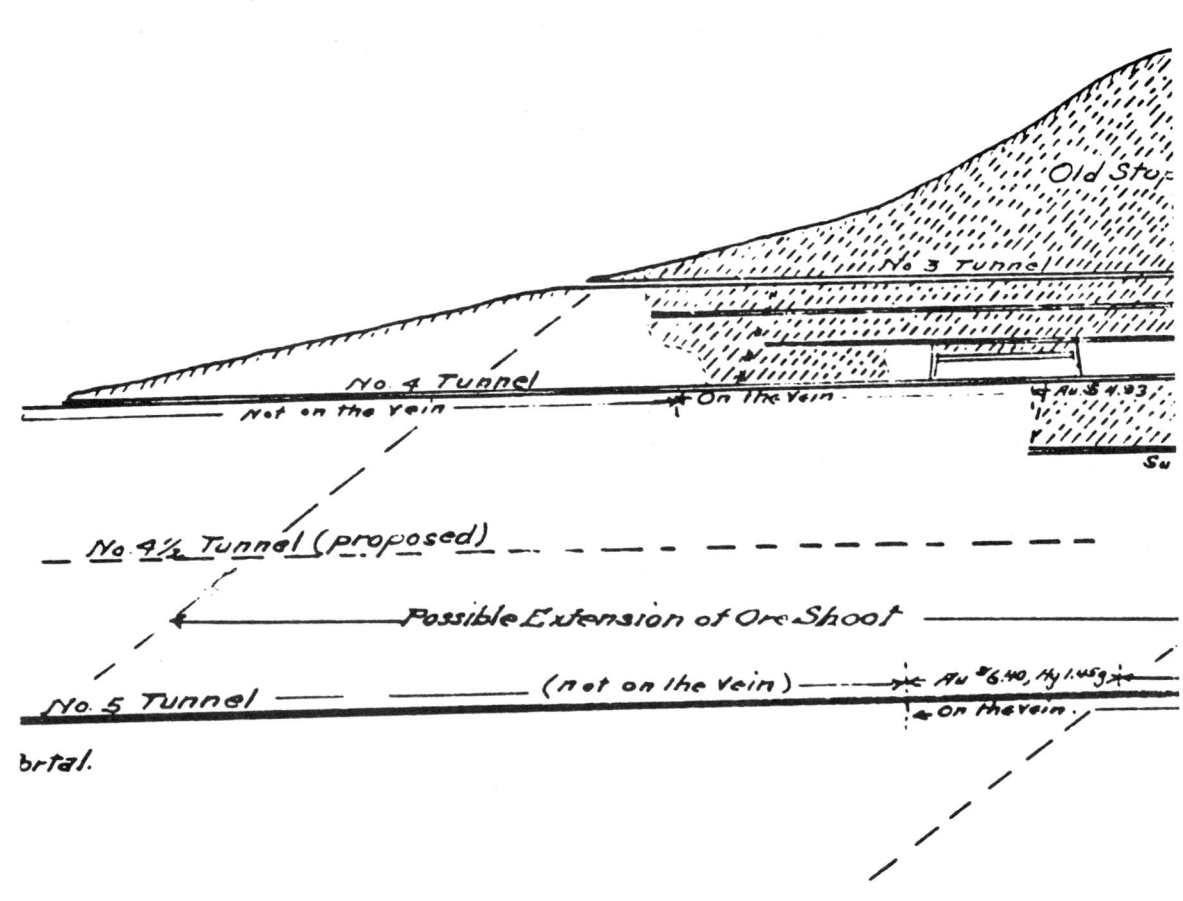

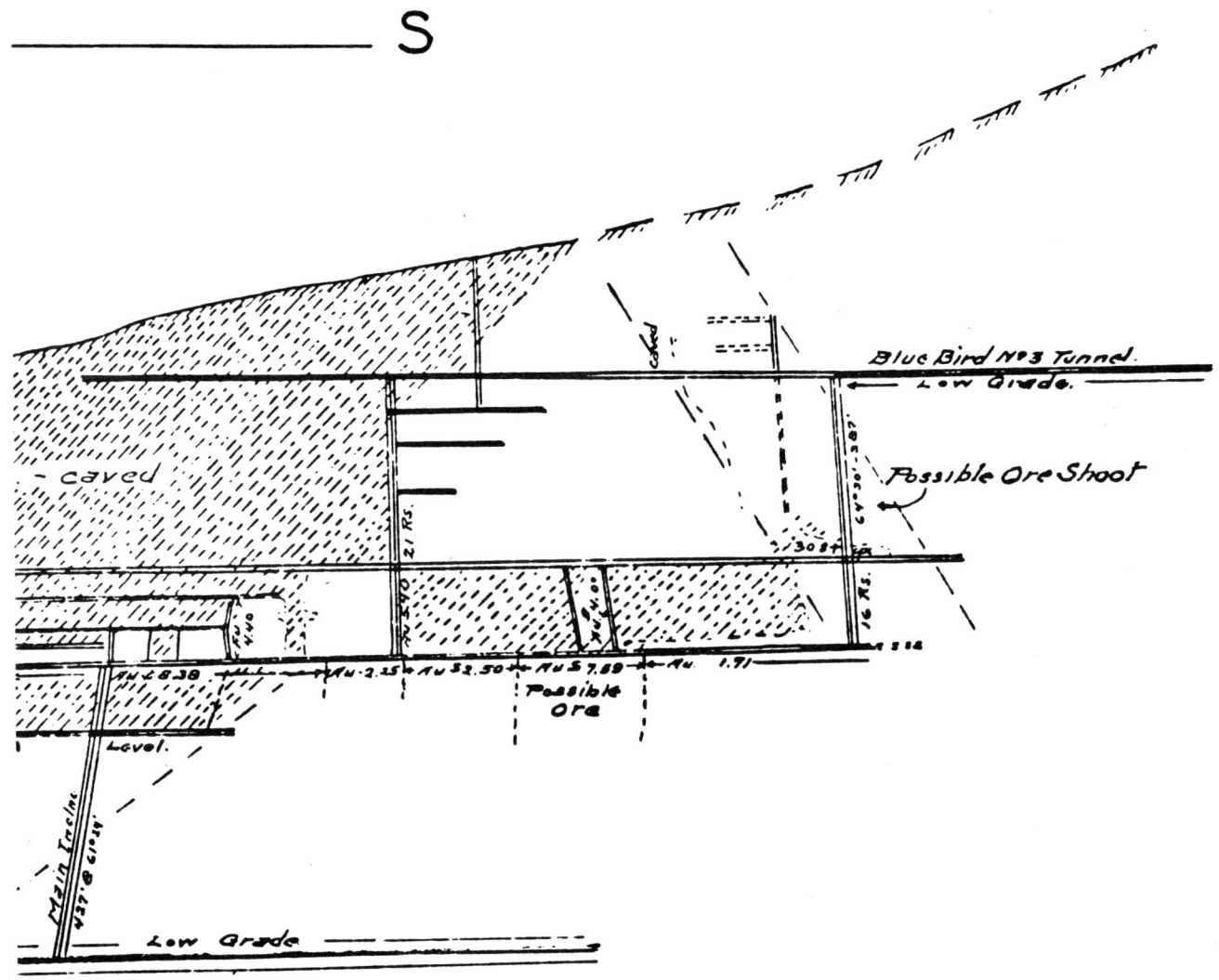

NORTH-SOUTH PROJECTION
ANNIE LAURIE MINE
DATA TAKEN FROM OLD MAPS AND REPORTS
SCALE 1" — 200'   OCT. 1930

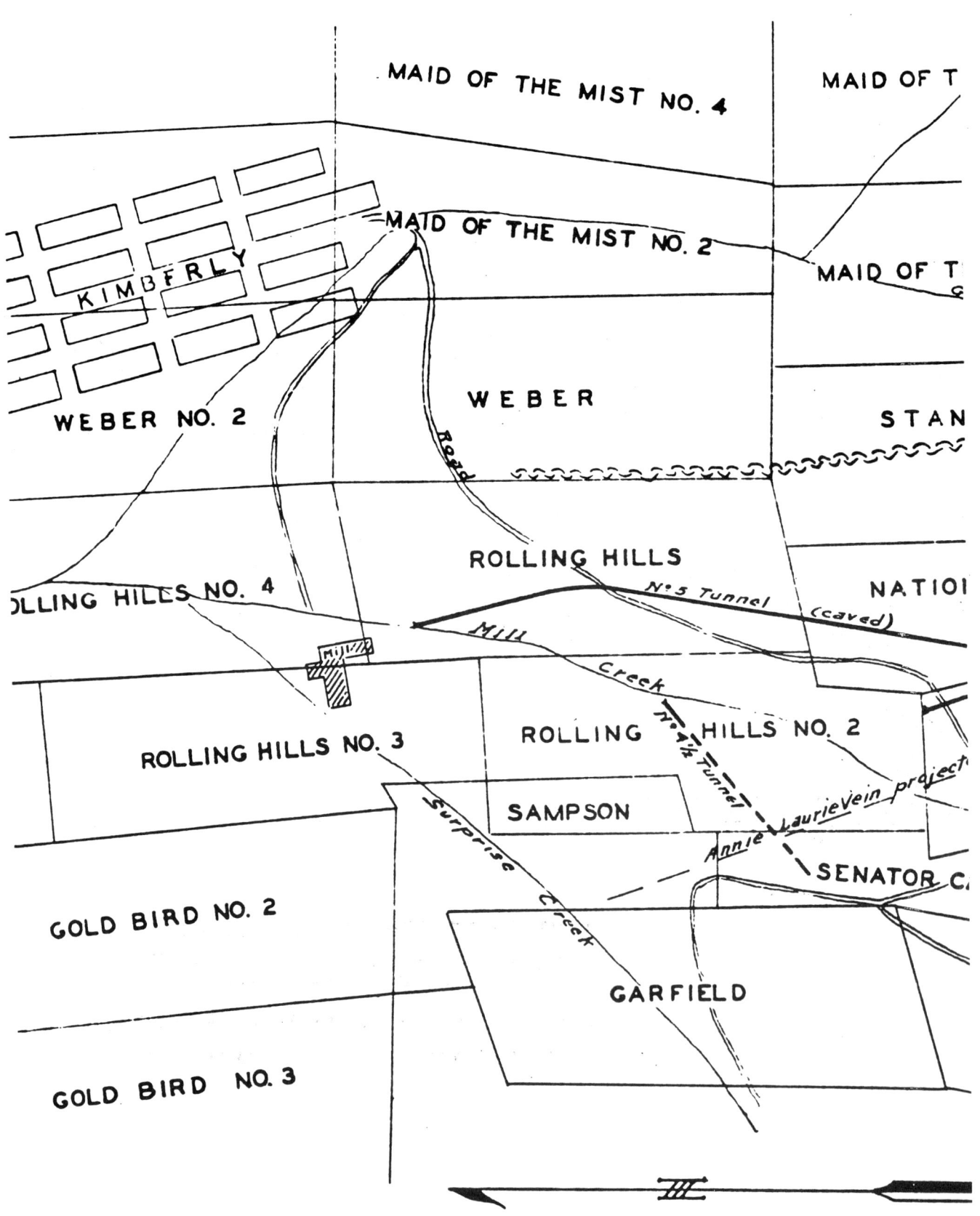

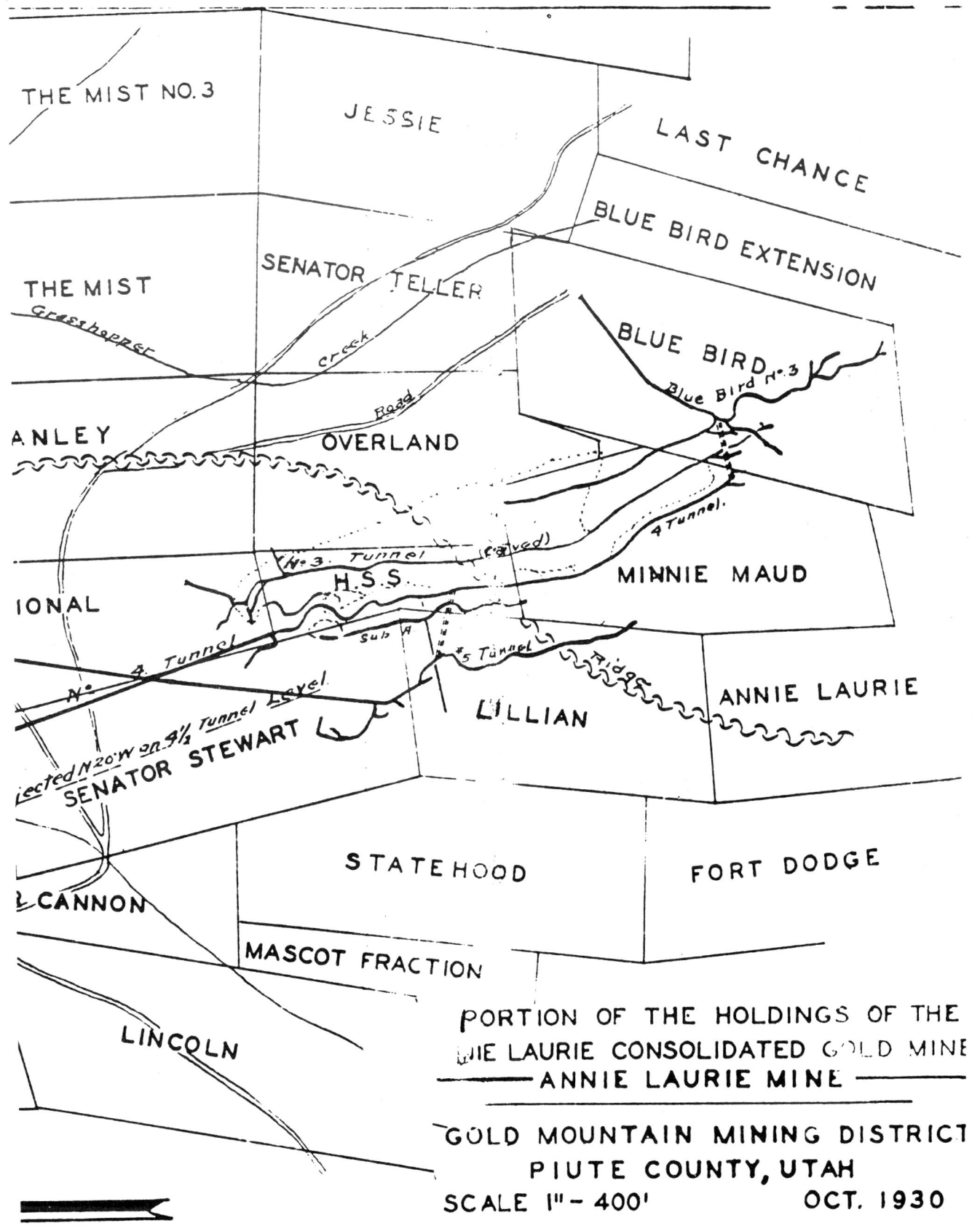

THE MIST NO. 3

JESSIE

LAST CHANCE

BLUE BIRD EXTENSION

THE MIST

SENATOR TELLER

Grasshopper

creek

BLUE BIRD *N°.3*

Blue Bird N°.3

ANLEY

OVERLAND

Road

4 Tunnel.

N°3 Tunnel (Faved)

H.S.S.

MINNIE MAUD

IONAL

N°. 4 Tunnel

Sub A

#5 Tunnel

Ridge

ANNIE LAURIE

Projected N20°W on 4½ Tunnel Level

SENATOR STEWART

LILLIAN

L CANNON

STATEHOOD

FORT DODGE

MASCOT FRACTION

LINCOLN

PORTION OF THE HOLDINGS OF THE
NIE LAURIE CONSOLIDATED GOLD MINE
———— ANNIE LAURIE MINE ————

GOLD MOUNTAIN MINING DISTRICT
PIUTE COUNTY, UTAH
SCALE I" – 400'   OCT. 1930

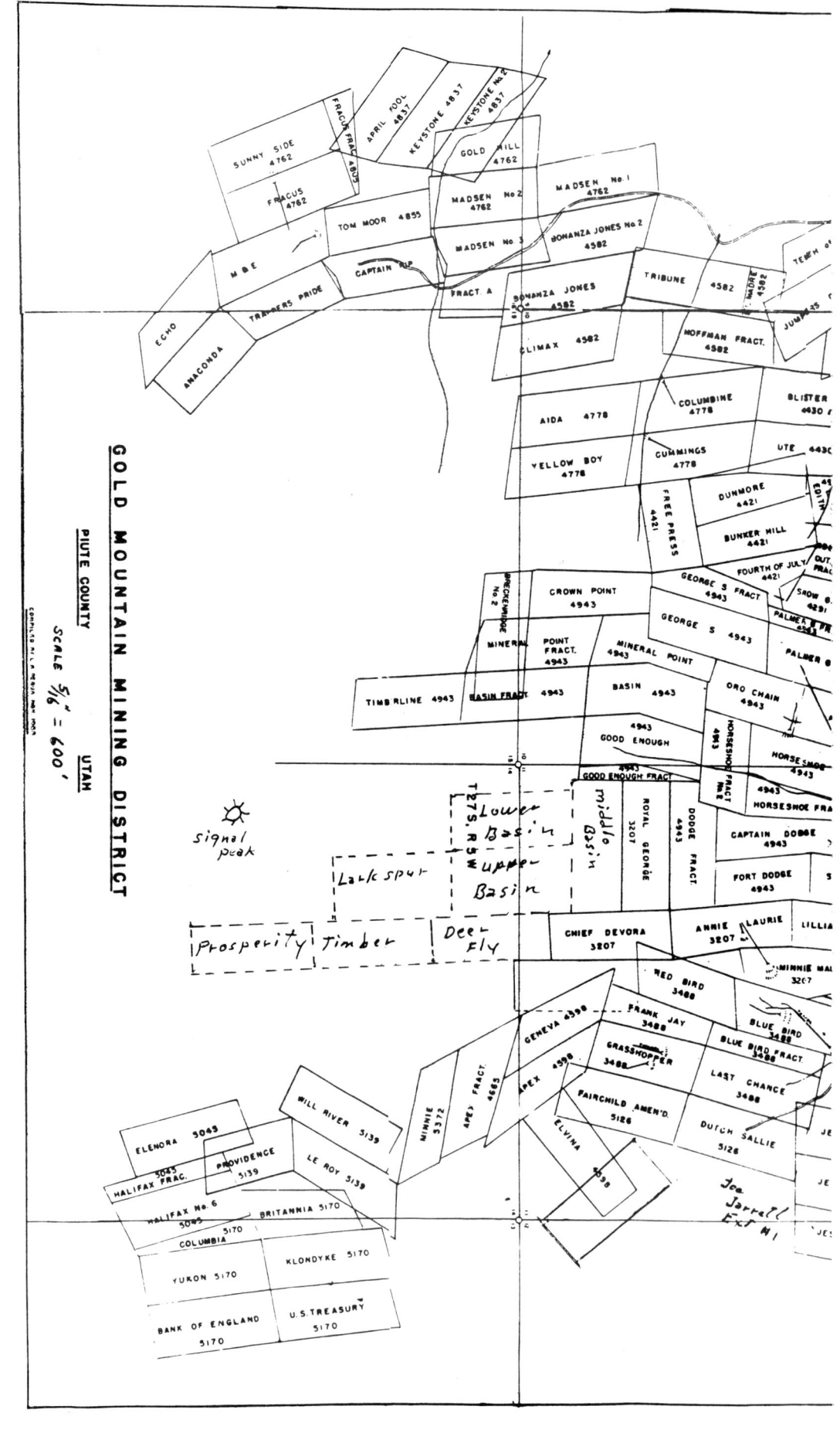

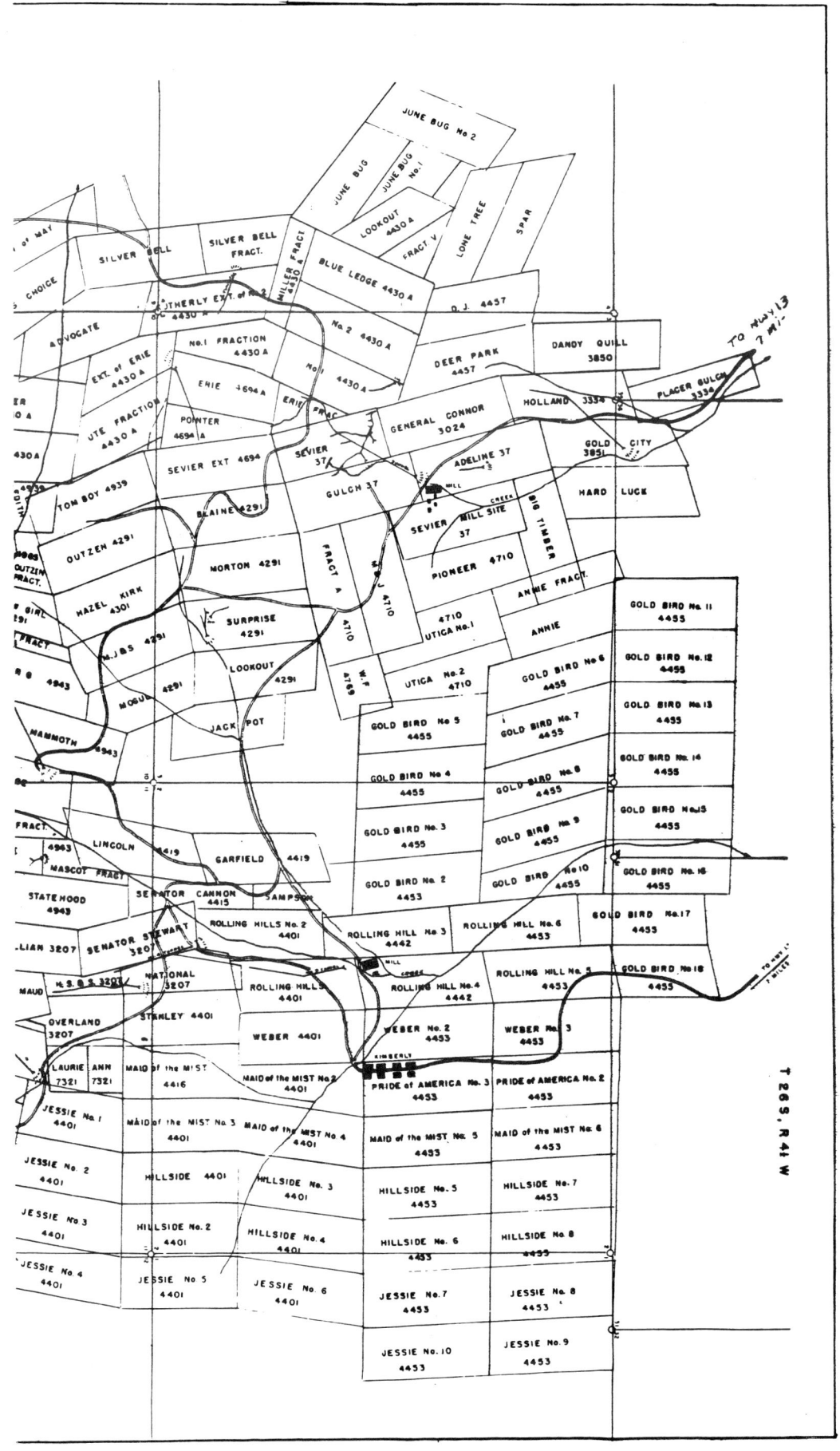

The following is a partial copy of the J. B. Wilson Report of January 28, 1915,
an evaluation of Kimberly's potential.

Report on the Annie Laurie Gold Mine
Kimberly, Piute County, Utah
Inspection and Report January 28, 1915

Location

The Annie Laurie Mine is located in a small glacial cirque near the top of the mountain range between Marysvale and Beaver. It is reached by a good wagon road from Sevier Station on the Marysvale Branch of the Denver and Rio Grande Railroad the distance being about 16 miles. The altitude of the mine is about 8,860 ft. while Sevier station is about 5,557 ft. above the sea.

History

The property was first operated extensively in 1899 when it was equipped with a mill and extensively developed. From September 30th, 1902, to April 22, 1905, it is said to have produced about $3,250,000 and to have paid $439,561 in dividends. At about this time trouble was experienced in holding the larger stopes open, probably owing to the system of cave mining which had supplanted the timbering of the stopes. This trouble increased until about 1906 when most of the mine had been lost and the Company had no funds with which to open more ground. The property then quietly passed through a receivership and operations were continued till 1907 when again the Company became insolvent and was sold at a receiver's sale. The purchaser operated the mine about two years under inefficient management and another receivership resulted in a sale to the present owners. The present management has worked the property continuously but in a small way doing development work and most of the ore now exposed in the workings is a result of their efforts.

It should be stated here that the reason for the failure of the original company can with much certainty be ascribed to mismanagement. The ore bodies are large and the hanging wall is none too solid and care and experience are necessary to hold the large stopes open. However, as long as the square set system was employed the property paid dividends, but eventually an attempt was made to work the stopes by a sub level method which I should judge would be hopelessly adapted to the conditions. The ground weakened by the numerous sub-levels that were run through the ore bodies and the miners who were employed in the workings at the time state that the ground began to move and the stopes were lost in rapid succession. The Company had paid out its earnings as dividends and was without funds to reopen the mine. The inefficient management covered their failure by announcing that the ore was exhausted and this has been generally believed

to be the case until more recent development work has disproved it.

Property

The properties owned by the present Company comprise what was formerly the Annie Laurie Mine, the Surprise MIne and the Sevier Mine. These cover a large number of patented claims as well as a number of unpatented locations. I did not see the Surprise nor the Sevier Mines. The former is unequipped but the latter is fully equipped and has a modern mill with water rights and a steam driven electric power plant. I believe the Sevier Mine has produced considerable ore but is irregular and much faulted and is said to look very unpromising. I did not visit it because I was told that neither it nor the Surprise Mine were included in the Annie Laurie Hholdings. However if the Annie Laurie Mine is to be investigated further these properties should be looked into with a view to including them in the deal if they should promise or if they have valuable equipment that could be used in the operation of the Annie Laurie Mine.

Geology

The ore occurs in a large fissure vein in Dacite and strikes North 20° West and dips about 65° West. The dacite seems to be a large volcanic stock or a deep intrusion that has become exposed by upheaval and erosion. The surrounding country is covered with rhyolite tuff which probably rests upon earl Tertiary sediments. The vein can be traced a long distance to the South but to the North a short distance is the rhyolite which might cut the vein off unless as is probably the case, the dacite underlies the rhyolite. The vein is developed in the workings for a length of about 2300 ft. and a depth of over 900 feet. The average thickness of the vein at the points sampled during this inspection is L2/4 ft. but in many of these sampled one or both walls were unexposed. The vein is very regular with well defined walls. The shoots are not limited by pinching of the vein but rather by the segregation of the values. I was not able to agree with the statement that there were two or more parallel veins in the workings but there are one or more veins branching off the main vein. This is certainly true of the Blue Bird Vein which strikes Northeast and Southwest and dips Northwest and intersects the Annie Laurie Vein in the Blue Bird No. 3 tunnel near the Shaw Raise. The Blue Bird Tunnels Nos. 1 and 2 are on this vein. The ore is white friable quartz with black manganese and red iron stains. Much of the gold is free and can frequently be seen in the ore.

Ore Bodies

Most of the ore produced by the mine came from a large shoot above the Annie Laurie No. 4 tunnel nearly 800 ft. in length, all of which is now caved. The ore which has been developed by the present management of the South of this large caved area. The men acquainted with the workings of the caved area maintain that a large percentage of the ore was lost in the old stopes and that much can now be receovered by raising into this ground. This seems to be true for a raise is now being carried into this ground and at the time of this inspection it was up over 100 feet and was in good ore. At this point the stoping had been confined to a slice one set wide along the hanging wall and the remainder of the vein seemed to be intact. Little difficulty was experienced in carrying this raise up and this makes it seem feasible to re-work at least some of this ground.

Further to the South the Shaw Rise, Williams Raise and the sub-levels above the Annie Laurie Tunnel No. 3 partly develop, say, 40,000 tons of ore which my random sampling indicates to carry Gold 0.228 oz., Silver 3.00 oz. With silver at 48 cents per ounce this gives the value of $7.20 per ton. This ore body is not developed on the No. 4 Tunnel but it seems likely that crosscutting at the south end of this tunnel should locate it which would give another 140 ft. of "backs" between Annie Laurie Tunnels No. 3 and No. 4. The North end of this ore body has not been reached but the management believes it will extend northward to the caved area a distance of probably 150 or 200 feet. Also to the south limit of this shoot is not proven conclusively by the workings and it seems probable that the ore will extend further in that direction.

The South workings in the Blue Bird Tunnel No. 3 and No. 4 do not show a very satisfactory condition. However, the extreme South end of Blue Bird Tunnel No. 3 follows a strong vein and a branch from this tunnel at this point follows another vein. There are good indications here that another vein has branched off similar to the condition presented by the Blue Bird Vein. Samples 10, 11, and 12 were taken in these workings in what seems to be a branch vein and sample No. 13 is on either the main vein or a crosscut vein. Undoubtedly ore of good milling grade can be develope din these workings. A small raise and stope here has produced ore formerly. The Annie Laurie No. 3 Tunnel seems to leave the main vein a short distance south of Stope #30 and angles off to the East. The crosscut at the south end of this tunnel should be carried further to the west where it should find the main vein. The South end of Annie Laurie Tunnel No. 4 is not far enough to the South to prove the extent of the ore developed in the upper workings. It also seems to have run off the ore near its South end. Sample No. 30 was taken in a crosscut in what appears to be the main vein

although this may be the Blue Bird vein branching off from the Annie Laurie Vein.

A sub-level was run about 90 feet below Tunnel No. 4 and extensive stoping is said to have been done from this sub-level. The Company records show this sub-level to have been about 500 feet long and to have sampled well throughout this distance. The No. 5 tunnel ran nearly 4,000 feet before reaching the vein which it approached as a quarter crosscut from the footwall side. The Maps show that this vein was drifted on for about 160 feet from the No. 5 Tunnel and the records show that the values were good throughout this distance. It is said however that the ground was "heavy." The No. 5 tunnel was equipped with electric haulage delivering the ore to the mill and was not apparently exploited extensively for ore as it was considered that the upper workings contained sufficient ore to render unnecessary immediate development elsewhere. This tunnel seems to have been very difficult to have kept open and probably for the reason that it ran under the bed of the gulch for most of its length and the gulch shows evidence of being due to erosion along a fault plain. It seems probable that with the reopening of No. 5 Tunnel a large reserve of ore should be developed between this tunnel and the sub-level below No. 4 Tunnel. Also it can reasonably be expected that this ore will extend further north than has been the case on the upper tunnels where it has been cut off by surface erosion and covered with glacial till.

There are other exposures of ore in the south workings which may develop important tonnages. One of these is between the Blue Bird Tunnels No. 2 and No. 3 in what has been called the Curnoe Raise.

Treatment

The mill was constructed before the treatment of slimes was perfected and it follows the usual method of the period of coarse crushing, drying the ore in tube roasters, crushing with fine rolls, classifying with impact screens, cyaniding by percolation and finally passing the tailings over amalgamation plates. Wood fuel was used for drying the ore. The mill contained 13 elevators and the capacity was seriously reduced by the usual elevator troubles. The rolls have been removed but the crushers, bins, cyanide tanks, piping and zinc boxes are in good condition and well housed. The cyanide tanks are of steel, on good foundations, and even the filter bottoms seem to be in good condition. There are ten 150-ton tanks and six 300-ton tanks. It is probable that if the mill were remodeled it would be equipped with stamps, amalgamation plates, classifiers and a slimes plant, and would use the present sand plant.

Appended to this report will be found a table showing the mill results from January 1st, 1901, to May, 1904.

*The following is a partial copy of the Earl B. Young Report of November 4, 1930
after a study of Kimberly's potential.*

Summary Report of Annie Laurie Mine Gold Mountain Mining District
Piute County, Utah
November 4, 1930

During the latter part of October, 1930, I spent several days at the Annie Laurie Mine situated in the Gold Mountain Mining District on the north slopes of the Tushar Mountains in Piute County, Utah. It is reached by auto road 16 miles from Sevier on the Marysvale Branch of the Denver and Rio Grande Railroad. The elevation is about 8,100 feet at the Annie Laurie Tunnel No. 5 and 8,900 feet at the Blue Bird No. 3 Tunnel.

Very complete reports were made on the property by Waldemar Lindgren in 1906, by G. B. Wilson in 1915, and H. E. Clement in 1922. As the mine is more completely caved than when these engineers reported, I am forced to rely for many facts on these early reports. Necessarily my report will be brief.

Summary and Conclusions

No. 1—The caved condition of the mine makes satisfactory independent examination impossible. Free use has been made of previous reports and maps, and data when they coincided with my impressions.

No. 2—The ore occurs in a fissure in dacite. The vein has been developed for about 2,400 feet in length and a depth of over 900 feet along the vein. Something like 450,000 tons of ore have been mined and dividends totaling nearly $440,000 have been declared.

No. 3—The management of the Annie Laurie Consolidated Mines proposes to drive Tunnel No. 4½ to cut the projected ore shoot of the Annie Laurie vein midway between No. 4 and No. 5 Tunnels. This development is recommended. At least $20,000 should be available for this work.

No. 4—When this development work opens up ore of tonnage and grade comparable with the possibilities, the mill should be redesigned and reconditioned probably on the basis of 200 tons a day. The cost probably would be not less than $75,000 using all available old buildings and equipment at the Annie Laurie and Sevier mills.

No. 5—It seems that an estimate of a possible tonnage of 400,000 tons above the old fifth level is justified. The grade of the ore is assumed to be 0.244 ozs. gold, and 3 ozs. silver, giving a head value of about $6 with 35¢ silver. Costs would probably be as follows: Mining $2, Development 40¢, Milling and Refining $1.30, General Expenses 40¢, Total $4.10.

No. 6—Assuming 90¢ extraction of $6 heads and allowing a cost of $4.10 gives a profit of $1.30 a ton. On 400,000 tons this would be $520,000, extending over a four to five year period at 200 tons per day. A larger mill would lead to greater profit in a shorter time, but capital investment will be larger.

Property

The Annie Laurie Mine is the property of the Annie Laurie Consolidated Gold Mines, a Utah corporation. The corporation is capitalied for three million shares, $1 par with 1,810,000 issued. The asked price of 75¢ would put a value of $1,357,500 on the property. This consists of 145 patented claims and 25 unpatented, the Annie Laurie Mine, the Sevier Mine, two cyanide mills, one of which is in good repair, an assay office, machine shop, office and store room, manager's residence, number of miners' cabins, also considerable equipment. An itemized list given in the report of H. E. Clement in 1922 gives a total cost of construction and equipment of over $480,000 at the Annie Laurie Mine. Much of this has been moved away or has deteriorated (and it is doubtful if what remains is worth much over one-tenth of the above figure. The mill and equipment at the Sevier Mine is in fair condition but no attempt was made to place a value on it. At the time of my visit, work was progressing on the transformer station which connects with the 44,000 volt line of the Telluride Power Company.

History

The earliest location is the Blue Bird in 1890. The Annie Laurie was operated extensively in 1899 when it was equipped with a mill. From 1901 to 1908 there was a production record of 425,000 tons and a bullion recovery of $2,785,000 or $6.555 per ton. Probably about a dollar of this was in silver with an average price of 57¢, the remainder being gold. From September 3, 1902 to April 22, 1905 dividends were declared amounting to $439,561. In 1908 the then operating company failed, a receiver was appointed and later the property passed to B. F. Bauer and associates. The present company was incorporated in 1930.

In 1913 considerable ore was developed at the south end of the mine above No. 4 Tunnel and nearly 10,000 tons were hauled and treated in the Sevier mill. The mill heads are said to have averaged 0.46 ozs. gold, and 3.14 ozs. silver, but it is claimed that this represented the richest part of the orebody then developed. Since that time work has been confined largely to maintaining No. 4 Tunnel which is now open to #16 raise—about one-half mile from the portal and to maintaining Blue Bird No. 3 Tunnel. No new orebodies or ore reserve have been opened or proved in recent years.

Workings

The Sevier Mine was developed by several tunnels and open cuts which are now caved. I was informed that on the lowest or No. 3 Tunnel level no ore was encountered. (The conclusion is made that the orebodies would not persist in depth.)

The Annie Laurie Mine was also developed by a number of long tunnels called the Blue Bird No. 3 and the Annie Laurie Nos. 3, 4, and 5. The vertical distance between Blue Bird No. 3 and Annie Laurie No. 5 is 800 feet. No. 4 is closely timbered and lagged from the portal to No. 16 raise over a half mile. The No. 5, the mill level tunnel, is caved and all workings below No. 4 Tunnel are full of water. All workings above No. 4 Tunnel are caved except No. 3 of the Blue Bird which is 387 feet above measured on the dip of the vein and a small portion of No. 3 of the Annie Laurie, 135 feet above, which can be reached from the No. 3 Blue Bird workings. The two maps submitted, herewith, give an idea in plan and section of the Annie Laurie workings and the position of the major ore bodies. The large stopes are reported to have been lost by an unsuccessful attempt to work by a caving system.

Geology

Surface exposures are poor. The ore occurs in a large fissure vein in dacite which strikes about N 20° W and dips 65° to the West. The dacite is probably a large volcanic stock. The vein is developed in the workings for about 2400 feet along the strike and over a depth measured along the vein of over 900 feet. The quartz forms an almost continual sheet along the vein varying from three to 20 feet and evidently with branchings of the vein and the inclusion of horses of waste in various places. The average thickness has been calculated as about 12 feet. In places there are streaks of sooty manganese and of iron oxides. The ore is a white friable quartz. Evidently there was calcite in part of the deposit "which has been dissolved by surface water leaving a hackly or lamellar quartz or striking appearance." As a rule there is little visible gold. Quoting from Lindgren, "These veins are believed to have been deposited by hot springs in fissures as a last phase of volcanic activity after the close of eruptions and not far below the actual surface of that epoch." "Mine workings have not penetrated below the zone of oxidation."

Ore Shoots and Ore Reserves

The principal ore shoot which is about 1300 ft. long as shown on the North South projection pitches at about 40° to the North. A smaller ore body which is exposed on Blue Bird No. 3 and Annie Laurie No. 3 is believed to pitch to the South. The ore remaining between these two shoots is believed to be low grade. As it was not possible to get samples of the larger ore shoot its assay value is taken from the former reports. Lindgren says "between $7 and $8 per ton." Wilson gives bullion sales for "entire period of 40 months ending May 31, 1904—211,321 tons—$7.2642 per ton. Expense $4.3359—Profit $2.9283." Clement says "value of the ore mined from 1901 to 1908 figuring an 84% recovery on average bullion receipts of $6.90 per ton for 412,000

tons was $8.22 per ton—silver at 57¢ per oz."

The smaller ore shoot pitches southward and is of doubtful value to me. My average of 11 samples taken at such openings in the hangwall and footwall as were accessible gave only Au. .14 oz., and silver 1.50 oz. Wilson assumed Gold 0.228 oz., Silver 3 oz. and Clement Gold .26 and Silver 3 oz.

As the men who made the earlier reports could see more of the mine than I could and as several of my samples had to be taken either in the hangwall or footwall of the zone and are hardly representative, I am willing to take an average of former analyses giving 0.244 ozs. gold and 3.0 ozs. silver, as probable values. At 35¢ silver this gives a value of $5.93 per ton. $6 is used for the large north dipping shoot.

Total tonnage is also difficult to estimate because of the caved condition of the property. Wilson allowed 40,000 tons south of the caved area and mentions other possible blocks for which no tonnage is estimated. Clement calculates 105,000 reasonably assured ore, 123,000 probable ore and 184,000 possible ore, total 412,000 tons with an assumed recovered value of $5.84 @ 60¢ silver. Recovery of gold estimated at 95% and silver at 50%.

Assuming for the south dipping block, 200 ft. long by 400 ft. high and an average width of 12 ft. and 14 cubic ft. to the ton, we get 70,000 tons of possible ore. Assuming for the large north pitching block between the four and five tunnels a length of 1300 ft. and a width of 10 ft. we get about 325,000 tons of possible ore. Also, it seems probable that some ore of mill grade can be recovered from the caved area although no estimate is possible here.

There is a possibility of other ore shoots both north and south of those shown on the maps. Probability of the extension of the ore zone below the No. 5 Tunnel also offers a hope for additional reserve.

Milling

The following is taken largely from the report by Mr. Clement. An analysis of the ore is as follows: Silica 93.65%; Ferric Oxide 1.60%; Calcium Oxide, 1.55%; Combined Water 2.20%; Sulphur, Nil; Manganese Oxide .70%; Total 99.70%.

"The gold is probably the residue from decomposition of auriferous pyrite, and the silver a sulphide in combination with the manganese minerals as well as alloyed with the gold." Proportion of gold to silver by weight is about 1 to 10. In 1907 the process was course crushing in No. 6 and No. 4 Gates gyratory crushers, drying in 5-19' revolving dryers using wood fuel, fine crushing to 35 mesh by rolls in circuit with impact screens—finished product conveyed to mill bins, then trammed by hand to cyanide vats. Cyanide solution of 8-10 pounds per ton of solution was used for 8 to 10 days; a 1/2 strength then used, and the charge then washed and sluiced to the plate room where copper plates were used for amalgamation. 93.4% of the gold was recovered, and 40% of the silver.

INDEX